Key Nuclear Reaction Experiments

Discoveries and consequences

Key Nuclear Reaction Experiments

Discoveries and consequences

Hans Paetz gen. Schieck

*Institute of Nuclear Physics,
University of Cologne, Germany*

IOP Publishing, Bristol, UK

ISBN 978-0-7503-1173-1 (ebook)
ISBN 978-0-7503-1174-8 (print)
ISBN 978-0-7503-1175-5 (mobi)

DOI 10.1088/978-0-7503-1173-1

Version: 20151001

IOP Expanding Physics
ISSN 2053-2563 (online)
ISSN 2054-7315 (print)

British Library Cataloguing-in-Publication Data: A catalogue record for this book is available from the British Library.

Published by IOP Publishing, wholly owned by The Institute of Physics, London

IOP Publishing, Temple Circus, Temple Way, Bristol, BS1 6HG, UK

US Office: IOP Publishing, Inc., 190 North Independence Mall West, Suite 601, Philadelphia, PA 19106, USA

To my wife Sybille who patiently supported me during the writing of this and other work.

Contents

Preface

With the 'age' of nuclear physics reaching 100 years it seems appropriate to consider in detail the historical aspects of how new knowledge has appeared. It is fascinating to look at theoretical developments in the field of subatomic physics in step with the grand new ideas of the 20th century, such as quantum theory and the theory of relativity, opening entirely new perspectives on the world. Experimental progress has given hints of 'new physics' (e.g. the stepwise deciphering of the structure and composition of nuclei led to the idea of two new interactions, the weak and the strong force and their theoretical description) and experiments were needed to decide between alternative theoretical interpretations (e.g. 'Are neutrinos Dirac or Majorana particles?', 'Do neutrinos have mass or not?' or 'Are electrons, neutrinos, muons and protons point particles or extended objects?').

In this respect it is worth studying the often fascinating details of early (in particular 'first') experiments. Quite often epoch-making results were obtained with very simple means, e.g. the discovery of nuclear fission. But all these experiments were based on earlier attempts that were carefully refined to yield unambiguous evidence, not pure serendipity (e.g. the carefully planned layout of the experiment to 'find' the neutrino).

This book is designed to give an outline of the key experiments in nuclear reactions. It is also motivated by an earlier book in German that I have often found very useful myself (Bodenstedt E 1972 *Experimente der Kernphysik und ihre Deutung* (Mannheim: BI Wissenschaftsverlag)), which comprises three volumes and also covers many nuclear structure experiments. The current book is restricted to nuclear reactions which seems justified, also because of the increased specialization of the subfields of nuclear physics.

In this book the 'crucial' or 'key' experiments that often, but not always, have been the first in a subfield are described in some detail, often including original drawings or set-ups because these may illustrate the igniting idea of a new field better than later and more sophisticated set-ups. Nevertheless in many instances later progress is briefly described. The theoretical background is given, but is kept compact and, if necessary, the usual textbooks or original literature will have to be consulted. Therefore, at the end of the introduction chapter references for general reading and other useful works are listed. Insisting on the reproduction of original drawings would result in reduced quality of the figures in some cases—therefore for these the figures have been redrawn or the text has been replaced.

About the author

Hans Paetz gen. Schieck

Born 1938 in Coburg, Germany. Studies of Physics, Mathematics, and Philosophy at Universities at Stuttgart, Hamburg, and Basel. Physics Diploma (1964) and Dr. phil. (1966) from the University of Basel. 1967–1970 Postdoc at Basel and Ohio State University, Columbus, OH. 1970 Visiting Assistant Professor at OSU. Since 1971 at the University of Cologne. Habilitation 1973. Apl. Professor 1978, University Professor since 1983. Retired since 2004. Member of the German Physical Society DPG, member and fellow of the American Physical Society APS. Author of books and many scientific articles in refereed journals.

Main fields of research: Low-energy nuclear reactions and particle spectroscopy. optical-model and Ericson fluctuation studies. Isobaric-analog studies in medium and heavy nuclei with polarized protons. Studies of few-body and fusion reactions in elastic scattering and breakup situations and search for three-body forces in comparison with Faddeev and EFT theories. Fusion reactions with respect to fusion-energy, especially study of the effects of polarization on the yield of fusion reactions. Spin physics and polarization methods: Polarized ^{3}He target, development of polarized ion sources of the atomic-beam and Lambshift type at Ohio State University, at Cologne, and for COSY (Jülich). Development of different polarimeters for protons and deuterons at Cologne and for COSY (Jülich), among them a unique Lambshift polarimeter.

IOP Publishing

Key Nuclear Reaction Experiments
Discoveries and consequences
Hans Paetz gen. Schieck

Chapter 1

Introduction

Key experiments are those which open up entirely new insights into unknown 'territories' and start new fields of more detailed investigations in these areas. One indicator of key experiments can be the awarding of Nobel Prizes to the principal investigators (examples are Robert Hofstadter (1961), and Jerome Isaac Friedman, Henry Way Kendall and Richard Edward Taylor (1990), for the study of the external and internal structure, respectively, of the proton by electron scattering).

In nuclear physics, a field of science which, by definition, has existed for only around 100 years, these key developments are *radioactivity* and the active investigation of nuclei, either their *structure* or their interactions in *nuclear reactions*. Both are intimately connected with the continuing progress in the development of particle accelerators and, in part, nuclear reactors.

1.1 Rutherford and evidence for the nuclear atom

Around 1911 in Manchester, UK, the famous Rutherford scattering experiment initiated the study of nuclear reactions. The behavior of α particles elastically scattered from gold and other nuclei suggested a very small (from the point of view of that time) and compact nucleus (i.e. containing most of the atom's mass and all of the positive charge Ze compensating that of the atomic electrons). Energetic particles from radioactive sources were used as projectiles (α particles from heavy elements such as 'radium emanation' ($^{222}_{86}$Rn) with sufficiently high energies and intensities). Even then scattering experiments were tedious: a MBq (in a 4π solid angle) source corresponds to an incident 'beam' current into a solid angle, small enough to define a reasonable scattering geometry, of only $\approx 1 \cdot 10^{-6}$ nA. Single scintillation events had to be counted by observing them on a ZnS screen in the dark.

1.2 The first true nuclear reaction

Around 1917 Ernest Rutherford, using techniques similar to those of the famous scattering experiment, recognized that a different type of particle emerged from the interaction of α particles from radioactive sources with gas molecules. This new particle had a longer range in matter than the α and proved to be the nucleus of the hydrogen atom. Rutherford had therefore performed the first true nuclear reaction with a rearrangement of the particles involved. Although the existence of negatively charged constituents (the atomic electrons) and positive ions had been seen in gas discharges, only Rutherford identified the particle which emerged from the

$$^{14}\mathrm{N} + \alpha \rightarrow {}^{17}\mathrm{O} + \mathrm{p} \tag{1.1}$$

reaction as the very small nucleus of H and coined the term 'proton'. Thus, he solved one part of the riddle of the structure and composition of nuclei; the other had to wait until the discovery of the neutron.

The nuclear charge number is identical to the element number of the periodic table and the Z dependence of the Rutherford cross section confirmed the periodic system of the elements. The explanation of the existence of isotopes and their correct placement in the chart of nuclides (Z versus N) required the discovery of the neutron in 1932. Rutherford could already—by comparing the measured scattering angular distribution of α particles on gold with his ansatz of a point-Coulomb interaction—conclude that the nucleus is an object smaller than the scattering distances (order of magnitude: $1\ \mathrm{fm} = 1 \cdot 10^{-15}\ \mathrm{m}$). The very fact that scattering at backward angles occurred, showed that the scattering center had to be heavier than the α (this is pure kinematics). The electron cloud relative to this is very large (order-of-magnitude radius: $1\ \text{Å} = 1 \cdot 10^{-10}\ \mathrm{m}$) and carries the charge $-Ze$ such that the atom is exactly neutral.

After the invention of accelerators the use of α particles of much higher energies with penetration into the target nucleus was possible and the extension (the radius) of nuclei could be obtained by the onset of deviations from the point-Coulomb scattering. A key role is played here by the *charge form factor* and its Fourier transform, the *charge density distribution*. It expresses how strongly the Coulomb potential of an extended (often simply assumed to be homogeneous) charge distribution in the nuclear interior deviates from that of a point charge or what the influence of the (hadronic) nuclear interaction on the observables is, see figure 4.3.

Using charged leptons, which have no measurable extension and do not feel the strong interaction, as probes, charge (and current) distributions in nuclei and nucleons have been determined. At higher momentum transfer (i.e. at high energies and large scattering angles via *inelastic or quasi-elastic scattering*) excited states of the nucleon and, later, (via deep-inelastic scattering) substructures of the nucleons (*partons*) were discovered which had all the properties of quarks—1/3 charges, spin $1/2\hbar$, color charge and confinement—characteristics of truly elementary particles (point shape, no internal structure) and, also by probing with neutrinos, they proved to be sources of the strong, electromagnetic and weak interactions.

1.3 The role of accelerators

It is evident that the use of radioactive sources imposed severe restrictions: a fixed or very limited energy range and extremely low intensities. It is clear that the field of nuclear reactions could only progress with the invention of particle accelerators. The first accelerator prototype important for nuclear physics was the linear accelerator (LINAC) developed and published in 1929 by Ralf Wideröe at the Aachen Institute of Technology, also laying the ground for the betatron, which was realized by Donald Kerst and Robert Serber in 1940, and the cyclotron by Ernest O Lawrence in 1931. Wideröe's ideas also included the synchrotron and storage ring schemes. The first nuclear reaction initiated with accelerated beams was the reaction

$$p + {}^{7}\mathrm{Li} \rightarrow 2\alpha \tag{1.2}$$

by John Cockroft and Ernest Walton in 1932 at the Cavendish Laboratory at Cambridge using a dc high voltage across several accelerating gaps and produced by the Delon/Greinacher voltage-multiplication scheme. This and the ensuing developments in nuclear and particle physics up to the present energies of up to 14 TeV (at the Large Hadron Collider at CERN/Geneva) are intimately connected with the achievements in accelerator physics and technology. Not unjustifiably, accelerators have been called 'tools of our culture' and 'engines of discovery' (see e.g. the book with that title by Sessler and Wilson [18]).

A very important part of this is the development of a variety of ion source types adapted to the special needs of different experiments. In many cases acceleration of negative ions is advantageous or required (e.g. in tandem Van de Graaff accelerators). The study of the spin dependence of nuclear reactions became possible with sophisticated sources of spin-polarized ions as well as spin-polarized targets. The possibility to produce and accelerate ions of very many isotopes is a prerequisite for, in particular, nuclear structure studies and, using exotic (e.g. radioactive) beams, the limits of the chart of nuclides are also being explored.

1.4 Detection methods

The fact that nuclear radiation cannot be seen (or felt) directly requires more or less sophisticated equipment to visualize or register the existence and interactions of different types of radiations such as α, β and γ particles, light and heavy ions up to fission product nuclei, neutrons and transuranium nuclei, etc. Thus, parallel to accelerator developments the development of detector technologies—from the first scintillators, later equipped with photomultipliers, to the cloud, spark and bubble chambers, the ionization chamber, the Geiger–Müller counter, multiwire ionization chambers and the large field of solid-state detectors—was essential. The impact of accelerators, especially in conjunction with modern detector technologies such as computed tomography for three-dimensional images, now extends into social applications such as tumor diagnosis and therapy, the identification and modification of materials, age and provenience analyses in archaeology, geology, arts, environmental science, security questions, etc.

1.5 The neutron and the correct composition of nuclei

With the detection of the *neutron* by James Chadwick (1932) (see also chapter 7) another branch of nuclear physics and, in particular, nuclear reactions opened up that only partly depends on accelerators. Not only was the discovery of the neutron the key to the fundamental structure of nuclei, removing all kinds of inconsistencies about, e.g., nuclear isotopes, but it also immediately incited Heisenberg to formulate the idea of charge independence of the nuclear interaction and the fundamental symmetry of *isospin*.

The neutrality of the neutron facilitates the description and also the execution of nuclear reactions. On the other hand, the production of neutrons for nuclear reactions as well as the detection methods are more complicated. Normally, except when neutrons from nuclear reactions are used, the choice or selection of specific neutron energies requires additional methods such as moderation by elastic collisions with light nuclei and/or chopper and time-of-flight facilities.

Much of the work on neutrons relies on neutrons from fission in reactors (an example is the high-flux 660 MW research reactor with a thermal flux of $>1 \cdot 10^{15}\ s^{-1}\ cm^{-2}$, at the Institut Laue-Langevin (ILL Grenoble)) or on spallation neutron sources where intense proton beams in the GeV and mA range incident on (liquid) metal targets release many (up to 30) neutrons per proton with high energies. A typical research center is the LANSCE facility with a proton LINAC, originally designed as a meson factory at Los Alamos, New Mexico, another is the spallation neutron source (SNS) at Oak Ridge, Tennessee, with 1.4 MW beam power and $4.8 \cdot 10^{16}$ neutrons s^{-1}.

The neutron has fundamental properties in its own right which have been studied:

- β decay.
- The internal (quark + gluon) structure and charge and magnetic moment distributions. These have been studied, e.g., using elastic and inelastic electron scattering where deuterons and, in particular, ^{3}He served as the neutron targets. Polarized ^{3}He is an almost pure polarized neutron target. The charge and magnetic moment distributions inside the neutron are proof of its inner structure.
- The possible electric dipole moment and thus time reversal and parity violations were studied where the absence of the Coulomb force is experimentally advantageous.
- The wave nature of neutrons of low energies was studied in reflection, diffraction and interference experiments.
- Ultracold neutrons in particular offer many interesting properties and applications, e.g. their interaction with the gravitational field or the interaction of their magnetic moment with magnetic fields.

1.6 Nuclear spectroscopy

We define nuclear spectroscopy as the science of learning all about the properties of the thousands of nuclides, each with individual and also collective properties. Aside from early studies of radioactive decays, nuclear reactions have been the main tool to investigate the action of nuclear forces (in the sense of an interplay of the strong interaction proper, and the electromagnetic and the weak force). In high-density situations, e.g. in neutron stars, even the gravitational force enters the stage via the density dependence of the nuclear interactions. The aim of modern nuclear spectroscopy is now moving away from stable nuclei, from deformed highly excited nuclei with high angular momenta to the investigation of nuclei in the regions near the limits of known nuclei with either high neutron excess, high neutron deficiency, or the region of new elements, the superheavy nuclei. These can be characterized by their isospin $T = (N - Z)/A$.

1.7 Higher energies

With higher energies available nuclear reactions produced a wealth of phenomena in intermediate- and high-energy physics, such as the *particle zoo* with thousands of more or less short-lived particles that do not exist naturally, ordered by the quark model. The inner structure of the hadrons (mesons composed of a quark and an antiquark, baryons consisting of three constituent quarks) could be investigated, as well as the deeper role of old and new conservation laws or invariances. The nature of the different forces and their interactions via the exchange of bosons, and their ranges and strengths are manifested in reactions, as is the answer to the fundamental question of how particles acquire mass (via the Higgs field or Higgs boson).

1.8 General references and resources

The specific source literature connected with each key experiment and given in each chapter is supplemented by the (selected) references of more general interest and the resources included in the bibliography of this chapter.

Bibliography

[1] Brink D M and Satchler G R 1971 *Angular Momentum* (Oxford: Oxford University Press)
[2] Edmonds A R 1960 *Angular Momentum in Quantum Mechanics* (Princeton, NJ: Princeton University Press)
[3] Eidelman S *et al* 2004 (Particle Data Group) *Phys. Lett.* B **592** 1
[4] Goldberger M L and Watson K M 1964 *Collision Theory* (New York: Wiley)
[5] Joachain C 1983 *Quantum Collision Theory* 3rd edn (Amsterdam: North-Holland)
[6] Lorenz-Wirzba H, Schmalbrock P, Trautvetter H P, Wiescher M and Rolfs C 1979 *Nucl. Phys.* A **313** 346
[7] Marmier P and Sheldon E 1970 *Physics of Nuclei and Particles* vol 1 (New York: Academic) ch 11.2
[8] Mott N F and Massey H S W 1965 *The Theory of Atomic Collisions* (Oxford: Clarendon)
[9] Newton R G 1966 *Scattering Theory of Waves and Particles* (New York: McGraw-Hill)
[10] National Nuclear Data Center, EANDC, Nuclear Reactions, http://www.nndc.gov
[11] Paetz gen. Schieck H 2012 *Nuclear Physics with Polarized Particles* (*Lecture Notes in Physics* vol 842) (Heidelberg: Springer)
[12] Paetz gen. Schieck H 2014 *Nuclear Reactions—An Introduction* (*Lecture Notes in Physics* vol 882) (Heidelberg: Springer)
[13] Redder A, Becker H W, Lorenz-Wirzba H, Rolfs C, Schmalbrock P and Trautvetter H P 1982 *Z. Physik* A **305** 325
[14] Rodberg L S and Thaler R M 1967 *Introduction to the Quantum Theory of Scattering* (New York: Academic)

[15] Rolfs C and Rodney W S 1988 *Cauldrons in the Cosmos* (Chicago: University of Chicago Press)

[16] Particle Data Group 2008 Review of particle properties *Rev. Mod. Phys.* **80** 633

[17] Satchler G R 1990 *Introduction to Nuclear Reactions* 2nd edn (London: McMillan)

[18] Sessler A and Wilson E 2007 *Engines of Discovery—A Century of Particle Accelerators* (Singapore: World Scientific)

IOP Publishing

Chapter 2

Rutherford scattering and the atomic nucleus

We begin with a universal definition of the fundamental observable of nuclear reactions, the (differential) cross section that can be applied in classical as well as quantum-mechanical descriptions.

Definition of cross section: The (differential) cross section is the number of particles of a given type from a reaction which, per target atom and unit time, are scattered into the solid-angle element $d\Omega$ (formed by the angular interval $\theta \ldots \theta + d\theta$ and $\phi \ldots \phi + d\phi$), divided by the incident particle flux j (a current density!).

With no azimuthal dependence this definition yields the classical formula for the cross section. With the number of particles $j\, d\sigma = j \cdot 2\pi b\, db$ one obtains

$$\left(\frac{d\sigma}{d\Omega}\right)_{\text{class}} = \frac{2\pi b\, db}{2\pi \sin\theta\, d\theta} = \frac{b}{\sin\theta} \cdot \left|\frac{db}{d\theta}\right|. \tag{2.1}$$

$\theta(b)$, which contains the dynamics of the interaction (the dynamics) is called the *deflection function*. Knowledge of this determines the scattering completely.

The measurements by Hans Geiger and Ernest Marsden and the interpretation of their experiment by Rutherford [1, 2] constitute a milestone in our understanding of the structure of nature and herald the true beginning of 'nuclear physics'. Their apparatus included all the features of modern scattering experiments, as can be seen in figure 2.1.

2.1 Rutherford scattering cross section

For the derivation of the classical Rutherford scattering cross section we assume that:

- The projectile and the scattering center (target) are point particles (with Gauss's law it can be proved that this is also fulfilled for extended particles as long as the charge distribution is not touched upon).
- The target nucleus is infinitely heavy (i.e. the laboratory system coincides with the center-of-mass (c.m.) system).

doi:10.1088/978-0-7503-1173-1ch2

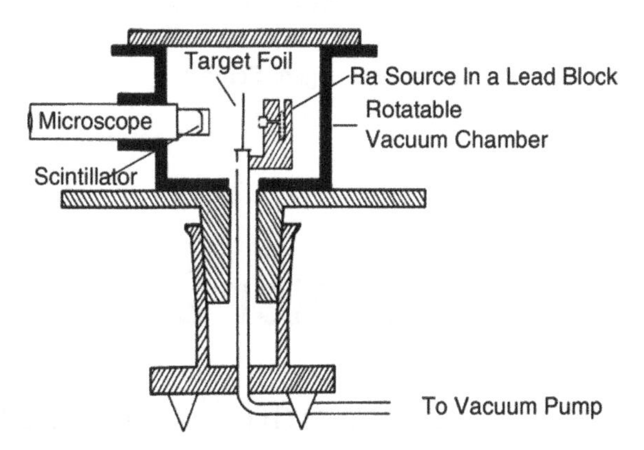

Figure 2.1. The original set-up of Rutherford, Geiger and Marsden's first nuclear scattering experiment in Manchester, 1908–13. Reproduced from [1], copyright 1913 Taylor and Francis.

- The interaction is the purely electrostatic point-Coulomb force.

$$F_C = \pm \frac{1}{4\pi\epsilon_0} \cdot \frac{Z_1 Z_2 e^2}{r^2} = \frac{C}{r^2} \tag{2.2}$$

  with the Coulomb potential $V_C = \pm C/r$.
- 'Classical' means that particles and trajectories are localized and no wave properties enter the description.

The classical scattering situation is shown in figure 2.2. The deflection function is most simply determined by applying angular-momentum conservation and the equation of motion in one coordinate (y):

$$L = mv_\infty b = mr^2\dot{\phi} = mv_{\min}\mathrm{d}, \tag{2.3}$$

(where v signifies a velocity) and from this

$$\mathrm{d}t = r^2\, \mathrm{d}\phi / v_\infty b \tag{2.4}$$

$$m\Delta v_y = \int F_y\, \mathrm{d}t$$

$$v_\infty \sin\theta = \frac{C}{mv_\infty b} \int_{-\infty}^{\infty} \dot{\phi} \sin\phi\, \mathrm{d}t$$

$$= \frac{C}{mv_\infty b} \int_0^{\pi-\theta} \sin\phi\, \mathrm{d}\phi = \frac{C}{mv_\infty b}(1 + \cos\theta). \tag{2.5}$$

After transformation to half the scattering angle the deflection function is

$$\cot(\theta/2) = mv_\infty^2 b/C = v_\infty L/C \tag{2.6}$$

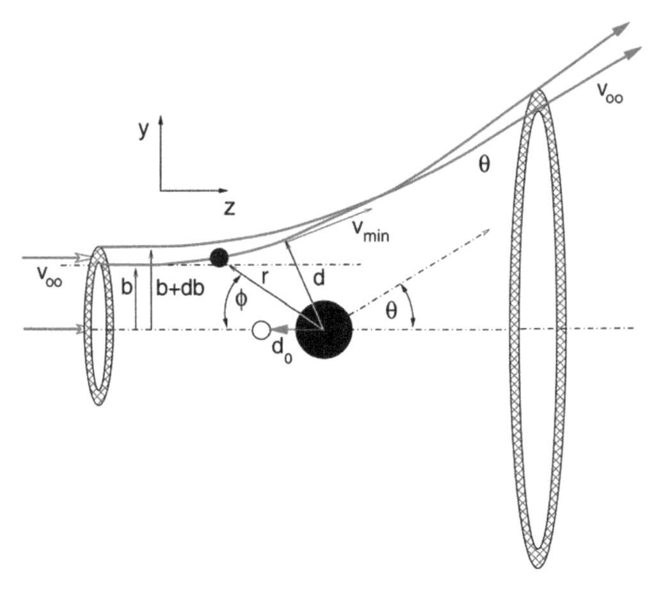

Figure 2.2. Classical Rutherford scattering. 'Classical' means that the particles involved as well as their trajectories are localized and there are no interference effects. Only the point-Coulomb force and classical conservation laws apply. With the proper definition of the cross section classical and quantum-mechanical predictions can be compared—it has to be considered as accidental that one obtain complete agreement between both.

and

$$b = \frac{C}{2E_\infty} \cdot \cot\left(\frac{\theta}{2}\right) \tag{2.7}$$

and

$$\frac{db}{d\theta} = \frac{C}{2mv_\infty^2} \cdot \frac{1}{\sin^2(\theta/2)} = \frac{C}{4E_\infty} \cdot \frac{1}{\sin^2(\theta/2)} \tag{2.8}$$

and thus for the Rutherford cross section

$$\frac{d\sigma}{d\Omega} = \frac{1}{(4\pi\epsilon_0)^2}\left(\frac{Z_1 Z_2 e^2}{4E_\infty}\right)^2 \cdot \frac{1}{\sin^4(\theta/2)}. \tag{2.9}$$

Numerically:

$$\frac{d\sigma}{d\Omega} = 1.296\left(\frac{Z_1 Z_2}{E_\infty\,(\mathrm{MeV})}\right)^2 \cdot \frac{1}{\sin^4(\theta/2)}\left[\frac{mb}{sr}\right]. \tag{2.10}$$

2.1.1 Minimal scattering distance *d*

For this quantity one needs additionally the energy conservation law:

$$\frac{mv_\infty^2}{2} = \frac{mv_{\min}^2}{2} + \frac{C}{d}. \tag{2.11}$$

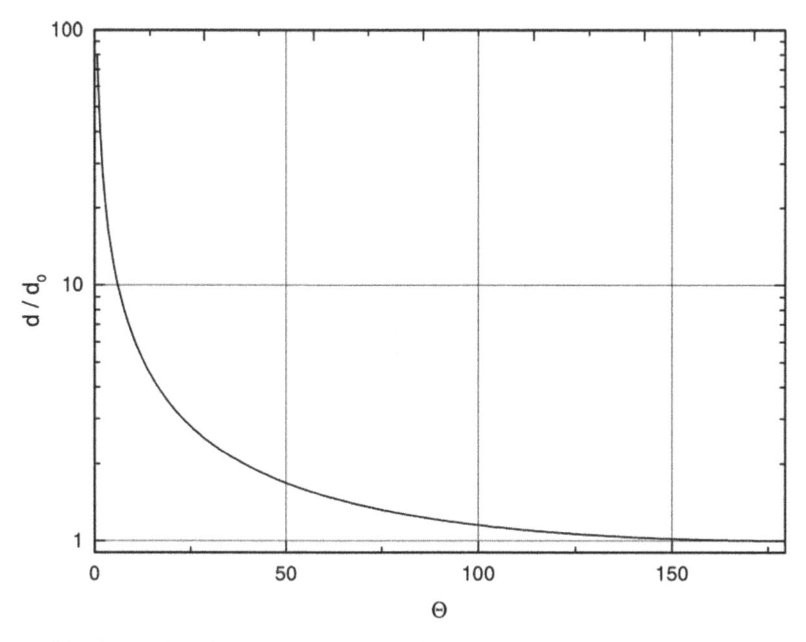

Figure 2.3. Minimal scattering distance as a function of the scattering angle. This classical quantity provides a feeling for how close nuclei as functions of energy and angle get in the scattering and—if one takes into account their radial extension—when they come into reach of the short-range nuclear force.

The absolutely smallest distance d_0 is obtained in central collisions with:

$$E_\infty = \frac{mv_\infty^2}{2} = \frac{C}{d_0}. \tag{2.12}$$

From this and the angular-momentum conservation (2.3) the relation

$$b^2 = d\left(d - d_0\right) \tag{2.13}$$

is obtained with the solution:

$$\begin{aligned} d &= \frac{C}{2E_\infty}\left(1 + \sqrt{1 + b^2\frac{4E_\infty^2}{C^2}}\right) \\ &= \frac{d_0}{2}\left(1 + \frac{1}{\sin\theta/2}\right). \end{aligned} \tag{2.14}$$

The classical scattering distance in relation to the minimum distance d_0 as a function of the scattering angle is shown in figure 2.3.

2.1.2 Trajectories in the point-charge Coulomb field

For the motion in a central-force field with a force $\propto r^{-2}$ classical mechanics shows that the trajectories for scattering, i.e. positive total energy, are hyperbolae, which can be derived using angular-momentum and energy conservation (with the Coulomb potential):

$$L = mr^2\dot{\phi} = \text{const} \tag{2.15}$$

$$E = \frac{mr^2}{2} + \frac{L^2}{2mr^2} + \frac{C}{r}. \tag{2.16}$$

In these equations dt can be eliminated. The integration of

$$\mathrm{d}\phi = -\frac{L}{mr^2}\left[\frac{2}{m}\left(E - \frac{C}{r} - \frac{L^2}{2mr^2}\right)\right]^{-1/2} \mathrm{d}r \tag{2.17}$$

results in

$$r = \frac{L^2}{mC} \cdot \frac{1}{1 - \epsilon \cos\phi} \tag{2.18}$$

with $b = L/\sqrt{2mE}$. With $k = L^2/mC$ and $\epsilon = \sqrt{1 + \frac{4E^2b^2}{C^2}}$ (the eccentricity) the standard form of conic sections is obtained

$$\frac{1}{r} = \frac{1}{k}(1 - \epsilon \cos\phi). \tag{2.19}$$

There is now a connection between impact parameter b, scattering angle θ, and (quantized) orbital angular momentum $L = \ell\hbar$

$$b = \frac{1}{2}d_0 \cot\frac{\theta}{2} = \frac{\ell\hbar}{p_\infty}. \tag{2.20}$$

2.1.3 Quantum-mechanical derivation of Rutherford's formula

The point-Rutherford cross section can be derived quantum-mechanically with identical results. This can be achieved in two ways. One is to solve the corresponding Schrödinger equation exactly, resulting in the regular and irregular Coulomb functions F_ℓ and G_ℓ as solutions. The other is to use the first Born approximation together with Fermi's golden rule.

Schrödinger equation

The point-Rutherford cross section may be derived quantum-mechanically by solving the Schrödinger equation with the point- (or extended-) Coulomb potential as the input. It has the form of a hypergeometric differential equation.

$$-\frac{\hbar^2}{2\mu}u'' + \left(\frac{C}{r} + \frac{\hbar^2}{2\mu}\frac{\ell(\ell+1)}{r^2} - \frac{\hbar^2k^2}{2\mu}\right)u_\ell = 0. \tag{2.21}$$

This equation may be written in its 'normal' form with the Sommerfeld parameter η_S and $\rho = kr$:

$$\frac{d^2u_\ell(\rho)}{d\rho^2} + \left(1 - \frac{\ell(\ell+1)}{\rho^2} - 2\frac{\eta_S}{\rho}\right)u_\ell(\rho) = 0. \tag{2.22}$$

It has the asymptotic solutions of the regular and irregular Coulomb functions with the Coulomb phases $\sigma_\ell = \arg\Gamma(\ell + 1 + i\eta_S)$:

$$F_\ell \longrightarrow \sin\left(kr - \ell\pi/2 - \eta_S \ln 2kr + \sigma_\ell\right), \tag{2.23}$$

$$G_\ell \longrightarrow \cos\left(kr - \ell\pi/2 - \eta_S \ln 2kr + \sigma_\ell\right). \tag{2.24}$$

With the usual partial-wave expansion with incident plane waves the Coulomb scattering amplitude of the outgoing wave results:

$$\Psi_S \longrightarrow \frac{1}{r}e^{i\left(kr-\eta_S \ln 2kr\right)}f_C(\theta), \tag{2.25}$$

$$f_C(\theta) = -\eta_S\frac{e^{2i\sigma_0} \cdot e^{i\eta_S \ln \sin^2\theta/2}}{2k^2\sin^2\theta/2}. \tag{2.26}$$

The amplitude squared $f_C \cdot f_C^*$ provides the Rutherford cross section, which is identical to the classically derived equation.

First Born approximation

The starting points for appropriate descriptions are:

- Fermi's golden rule of perturbation theory.
- The first Born approximation.

For a 'sufficiently weak' perturbation Fermi's golden rule gives the transition probability per unit time W:

$$\begin{aligned} W &= \frac{2\pi}{\hbar}\left|\left\langle\Psi_{out}\left|H_{int}\right|\Psi_{in}\right\rangle\right|^2\rho(E) \\ &= \frac{Vmp\,d\Omega}{4\pi^2\hbar^4}\cdot\left|H_{if}\right|^2. \end{aligned} \tag{2.27}$$

The density of final states $\rho(E) = dn/dE$, which enters the calculation can be obtained from the ratio of the actual to the minimally allowed phase-space volumes:

$$\frac{dn}{dE} = \frac{V4\pi p^2\,dp\frac{d\Omega}{4\pi}}{(2\pi\hbar)^3\,dE}, \tag{2.28}$$

$E = p^2/2m$ and $dp/dE = m/p = E/c^2p$. Thus

$$\rho(E) = \frac{dn}{dE} = V\frac{pm\ d\Omega}{(2\pi\hbar)^3}$$
$$= V\frac{pE\ d\Omega}{(2\pi\hbar)^3c^2}. \tag{2.29}$$

W becomes the cross section according to the definition on page 1 with the incident particle current density $j = v/V = p/mV$:

$$d\sigma = \frac{W}{j} = \frac{W}{\left(\frac{p}{mV}\right)} = \frac{V^2m^2\ d\Omega}{4\pi^2\hbar^4} \cdot \left|H_{if}\right|^2. \tag{2.30}$$

The first Born approximation consists in using only the first term of the Born series with plane waves in the entrance and exit channels:

$$\Phi_{\rm in} = \frac{1}{\sqrt{V}}e^{i\vec{k}_i\vec{r}} \qquad \text{and} \qquad \Phi_{\rm out} = \frac{1}{\sqrt{V}}e^{i\vec{k}_f\vec{r}}. \tag{2.31}$$

If $H_{\rm int} = U(r)$ signifies a small time-independent perturbation then, with $\vec{K} = \vec{k}_f - \vec{k}_i$

$$\left|H_{if}\right| = \left|\frac{1}{V}\int e^{i\vec{K}\vec{r}}U(r)d\tau\right| \tag{2.32}$$

and

$$\frac{d\sigma}{d\Omega} = \left(\frac{m}{2\pi\hbar^2}\right)^2 \left|\int e^{i\vec{K}\vec{r}}U(r)d\tau\right|^2 = |f(\theta)|^2. \tag{2.33}$$

Inserting the Coulomb potential $U(r) = C/r$ the classically calculated formula for the Rutherford scattering cross section is obtained. The cross section is (with the constant $Z_1Z_2e^2/16$ and the substitution $u = iKr\cos\theta$ and $du = -\sin\theta\ d\theta(iKr)$)

$$\frac{d\sigma}{d\Omega} = \text{const} \cdot \left|\int e^{i\vec{K}\vec{r}} \cdot \frac{1}{r}d\tau\right|^2$$
$$= \text{const} \cdot \left|\int\int \frac{1}{r}e^{iKr\cos\theta}2\pi\sin\theta\ d\theta\ r^2\ dr\right|^2$$
$$= \text{const} \cdot 2\pi\left|\int\int \frac{r}{iKr}e^u\ du\ dr\right|^2$$
$$= \text{const} \cdot \left(\frac{2\pi}{iK}\right)^2 \left|\int_r \left(e^{iKr\cos\pi} - e^{iKr\cos 0}\right)dr\right|^2$$
$$= \text{const} \cdot \left(\frac{2\pi}{iK}\right)^2 \left|\int_r \left(e^{-iKr} - e^{iKr}\right)dr\right|^2$$
$$= \text{const} \cdot \left(\frac{2\pi \cdot 2i}{iK}\right)^2 \left|\int_0^\infty \sin Kr\ dr\right|^2. \tag{2.34}$$

The integral is undefined. This is circumvented by a *screening ansatz* after Niels Bohr, which corresponds to the real situation of the screening of the point-Coulomb potential by the electrons of the atomic shell, with the screening constant α. With

$$\int_0^{\infty} e^{-\alpha r} \sin Kr \, dr = \frac{K}{K^2 + \alpha^2} \tag{2.35}$$

one obtains

$$\left(\frac{d\sigma}{d\Omega}\right)_{R,s} = \left[\frac{2\mu Z_1 Z_2 e^2}{\hbar^2\left(\alpha^2 + 4k^2 \sin^2(\theta/2)\right)}\right]^2 \tag{2.36}$$

with the momentum transfer $K = 2k \sin(\theta/2)$ for elastic scattering. This cross section is finite for $\theta \to 0°$. By letting the screening constant go to zero a cross section results, which is identical with that from the classical derivation:

$$\left(\frac{d\sigma}{d\Omega}\right)_{R} = \lim_{\alpha\to 0}\left(\frac{d\sigma}{d\Omega}\right)_{R,s}$$

$$= \left(\frac{Z_1 Z_2 e^2}{4E_{\mathrm{kin}}}\right)^2 \cdot \frac{1}{\sin^4(\theta/2)}. \tag{2.37}$$

However, for all applications where there is interference the Rutherford *amplitude* has to be used including its (logarithmic) phase. Typical cases are those of identical particles (see chapter 13) or of interference with nuclear (hadronic) amplitudes. Normally one has to assume a fundamentally quantum-mechanical description that may only be approximated by classical methods in special cases, i.e. when the relevant de Broglie wavelengths are small. For this decision the *Sommerfeld criterion* has been formulated.

$$\lambda\!\!\!^{-}_{\text{de Broglie}} = \hbar/p \ll d. \tag{2.38}$$

When choosing for a typical object dimension half the distance of the trajectory turning point d_0 for a central collision the Sommerfeld criterion for classical scattering is obtained

$$\eta_S = \frac{Z_1 Z_2 e^2}{\hbar v} = Z_1 Z_2 \frac{e^2}{\hbar c} \cdot \frac{c}{v} = Z_1 Z_2 \cdot \frac{\alpha}{\beta} \gg 1 \tag{2.39}$$

or numerically (for a heavy target)

$$\eta_S \approx 0.16 \cdot Z_1 Z_2 \sqrt{\frac{A_{\mathrm{proj}}}{E_{\mathrm{lab}}\ (\mathrm{MeV})}} \gg 1. \tag{2.40}$$

2.1.4 Results of the experiment

The results of the Rutherford–Geiger–Marsden experiment are shown in figure 2.4. The figure exhibits the strong angle dependence of this cross section together with the original data of [1], adjusted to the theoretical curve shown.

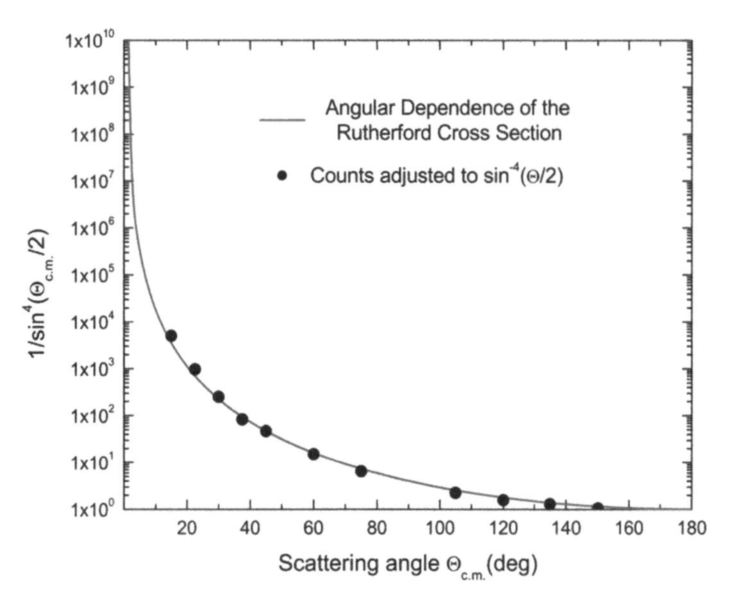

Figure 2.4. The curve shows the angular dependence of the theoretical Rutherford cross section $\propto \sin^{-4}(\theta/2)$. The points are the original data (which consisted of tabulated numbers of counts with no error bars, not transformed into cross-section values) of [1], adjusted to the theoretical curve, giving a nearly perfect fit (nowadays data with at least an error estimate or, better, error bars would be mandatory).

2.1.5 Consequences of the Rutherford experiments and their historic significance

Rutherford and his collaborators Geiger and Marsden (later also James Chadwick) used α particles from radioactive sources as projectiles. Their energies were so small that for all scattering angles the minimum scattering distances d were large compared with the sum of the two nuclear radii of the projectiles and targets. The complete agreement between the results of the measurements and the (point-) Rutherford scattering cross-section formula shows this in accordance with Gauss's law of electrostatics: a finite charge distribution in the external space beyond the charges cannot be distinguished from a point charge with an r^{-1} potential. In addition, the mere occurrence of backward-angle scattering events proves uniquely by simple kinematics that the target nuclei were heavier than the projectiles. Thus the existence of the atomic nucleus as a compact (i.e. very small and heavy object) was established (and Thomson's idea of a 'plum pudding' of negative charges from distributed electrons, in which the positive charges of ions were suspended, was refuted).

Later, the energy dependence as well as the dependence on charge numbers were fully corroborated leading to the confirmation of and a few corrections to the periodic table of elements.

Bibliography

[1] Geiger H and Marsden E 1913 *Phil. Mag.* **25** 604
[2] Rutherford E 1911 *Phil. Mag.* **21** 669

IOP Publishing

Key Nuclear Reaction Experiments
Discoveries and consequences
Hans Paetz gen. Schieck

Chapter 3

The first true nuclear reaction and the discovery of the proton

The first 'true' nuclear reaction (i.e. one with transmutation into different particles) was discovered by Rutherford in 1919—after earlier work with Ernest Marsden, in which particles with a larger range, ^{1}H nuclei, than that of the α particles in scattering from different targets were observed. In the crucial experiment a radioactive source with 6 MeV α was used. The impact of charged particles on a scintillating screen was observed through a microscope as a function of the distance from the source. The unknown radiation still appearing beyond the range of elastically scattered α consisted of 'protons'. The reaction leading to protons was identified as

$$\alpha + {}^{14}\mathrm{N} \rightarrow {}^{17}\mathrm{O} + \mathrm{p} \tag{3.1}$$

with a Q-value of −1.193 MeV. In the forward direction the initial energy of the protons was about 4.6 MeV corresponding to a mean range in air at NTP conditions of about 28 cm as compared to a range of the original αs of about 4.7 cm. The discovery of the proton took place in a simple chamber filled with air (i.e. mainly nitrogen) containing an α source at variable distances from the screen, see figure 3.1.

Figure 3.2 shows such a single event observed a few years later in a cloud chamber. It also shows an event of 'Rutherford' scattering of the recoil ^{17}O nucleus on an ^{14}N nucleus. The cloud chamber, which is still unsurpassed as an instrument for visualizing such events and also cosmic rays, etc, was invented by Charles Wilson after 1911, but only developed for practical use by Patrick Blackett after 1921, so was not yet in use by Rutherford. Consequently the ^{1}H nuclei were identified as part of all nuclei and Rutherford coined the term 'proton'. This term was proposed by Rutherford at a meeting of the British Association for the Advancement of Science at Cardiff in 1920 and found general approval by members of Section A there, according to a footnote by Rutherford himself to a paper of Orme Masson [5].

doi:10.1088/978-0-7503-1173-1ch3

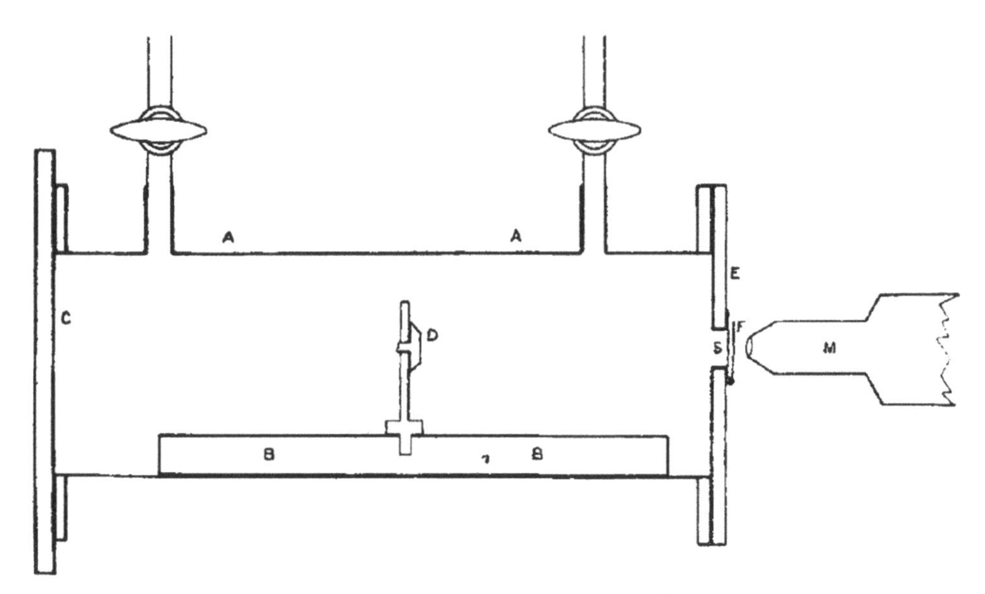

Figure 3.1. The apparatus used by Rutherford from 1917–20 to bombard ^{14}N with α particles. It consisted of a rectangular brass vessel 18 cm in length that could be evacuated and filled with different gases (hydrogen, dry air, nitrogen, or oxygen). D is a disk-shaped radioactive source on a sliding holder, S is the zinc sulphide scintillating screen and M is the microscope for observing the scintillations with a field of view of 2 mm. The emitted particle radiation of longer range was identified as consisting of $Z = 1$, $A = 1$ particles, forming the nucleus of the hydrogen atom, and for which Rutherford coined the word 'proton' in 1919. Reproduced from [9]. Copyright 1919 Taylor and Francis.

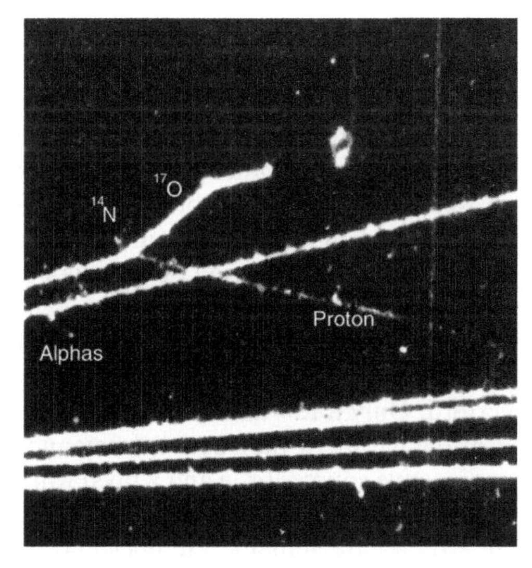

Figure 3.2. From a cloud chamber photograph (Plate 7, Photograph No. 1) by Blackett in 1925 [3] of the nuclear reaction $\alpha + {}^{14}\mathrm{N} \rightarrow {}^{17}\mathrm{O} + \mathrm{p}$, first observed by Rutherford in 1919 [9]. The figure shows not only the proton, emitted from the reaction with a ^{14}N nucleus, but also the recoiling ^{17}O, undergoing elastic scattering.

With the proton an essential component of nuclei had been found. With only α particles, electrons and protons as the known particles at the time, the properties of nuclei could not be explained satisfactorily. By comparison of the characteristic x-ray spectra of different elements with the charge number Z from Rutherford scattering it became clear that Z characterizes the chemical elements. Frederick Soddy found that in some cases the radioactive decays of chemically identical elements could be different and called these different substances *isotopes* [11], see also [4]. Already in 1919, with the development of the high-resolution mass spectrograph by Aston *et al* [1, 2], it became clear that almost all elements had a number of isotopes that could be separated according to their masses in the mass spectrograph. The atomic masses of all isotopes were nearly, but not exactly, integer multiples of the atomic mass of hydrogen, the small deviations being due to the binding energy of nuclei. The possible structure of atomic nuclei was difficult to explain: they could not consist of only protons and the partial compensation of the nuclear charge by electrons in the nuclei is excluded by quantum mechanics (which, however, was only developed after 1924). Rutherford postulated already in 1920 that there must be an additional particle with no charge and atomic mass number 1 which he christened the neutron [10]. However, the remaining puzzles about the true structure of nuclei could only be resolved after the neutron was discovered by James Chadwick, which was not until 1932 (see chapter 7). Others had mistakenly interpreted the neutron radiation from the reaction $\alpha + {}^{9}\mathrm{Be} \rightarrow {}^{12}\mathrm{C} + \mathrm{n}$ (with α from a polonium source) to be an energetic γ radiation.

The paper [9] in which Rutherford announced the first observation of a nuclear transmutation is part of a longer series on 'Collisions of α particles with light atoms' showing in an exemplary manner by many different experiments that the reaction particles were really hydrogen nuclei, that they were not produced as recoils from hydrogen in the source or target materials and that their energies (measured by their ranges in different materials) could only be explained by nuclear rearrangements (i.e. true nuclear reactions). Mass spectrometry through the application of electric and magnetic fields as well as studies of the brightness of the scintillations were performed. All this work required an unusual amount of patience, perseverance and precision because of the method of observation of the small and faint scintillations through a microscope.

The detection of nuclear reactions and the identification of protons as constituents of nuclei determining the nuclear charge are but two of the many remarkable achievements of Rutherford. Only a few additional ones will be listed here:

- The explanation of the nature of radioactive decay and the different types of radiation from radioactive sources (α, β and γ rays) which won him the Nobel Prize in chemistry in 1908 [6].
- The proof that α particles are actually ${}^{4}\mathrm{He}$ nuclei [7].
- The proof that atoms have a dense and massive core, a true *nucleus* that also became the basis of Niels Bohr's atomic model [8].
- The postulation of a possible neutral constituent of nuclei, the neutron [5].

With the advent of accelerators in particular (see chapter 9), which allowed one to change the energies of incident projectiles as well as providing a more or less free choice of beam and target particle combinations, the field of nuclear reactions (including elastic, inelastic and rearrangement processes) developed rapidly. The reaction mechanisms under the influence of the three fundamental interactions offered a rich field of study. The investigation of the rich structure of the numerous (>3300) stable and radioactive nuclides relies heavily on nuclear reactions to excite their quantum states or open up all kinds of reaction channels leading partly to very exotic nuclei. An example of a special scientific application is the reaction chains of nuclear astrophysics trying to describe the pathways from nucleosynthesis shortly after the Big Bang by fusion reactions between light nuclei up to the mechanisms leading to the heavy (and superheavy) species. Such processes still take place in the formation of new stars in the Universe.

Bibliography

[1] Aston F W 1919 *Phil. Mag.* **38** 709
[2] Aston F W 1920 *Phil. Mag.* **39** 449
[3] Blackett P M S 1925 *Proc. R. Soc.* A **107** 349
[4] Fajans K 1913 *Phys. Z.* **14** 131, 136
[5] Masson O 1921 *Phil. Mag.* **41** 281
[6] Rutherford E and Soddy F 1902 *Phil. Mag.* **4** 370, 569
[7] Rutherford E and Geiger H 1908 *Proc. R. Soc.* A **81** 162
[8] Rutherford E 1911 *Phil. Mag.* **21** 669
[9] Rutherford E 1919 *Phil. Mag.* **37** 537
[10] Rutherford E 1920 *Proc. R. Soc.* A **97** 374
[11] Soddy F 1913 *Chem. News* **107** 97

IOP Publishing

Key Nuclear Reaction Experiments
Discoveries and consequences
Hans Paetz gen. Schieck

Chapter 4

Extended matter and charge distributions of nuclei

Naturally the study of the finite size of nuclei requires higher-energy projectiles, as has been indicated by classical arguments in chapter 2. These can be hadronic particles such as α with medium energies (around 50 MeV would be sufficient due to the small de Broglie wavelength) or, e.g., electrons with energies of hundreds of MeV. Because interference effects occur between the strong and Coulomb interactions, and electrons are relativistic, a classical description is impossible.

The extension of the derivation of the Rutherford cross section to an extended (especially a homogeneous and spherically symmetric) charge distribution is simple and leads to the fundamental concept of the form factor.

We start with the Coulomb potential of such an extended homogeneous spherical charge distribution (figure 4.1) as a model distribution. It is calculated with Gauss's theorem of electrostatics:

$$V(r) = \begin{cases} \frac{Ze^2}{4\pi\epsilon_0}\frac{1}{r} & \text{for } r > R \\ \frac{Ze^2}{4\pi\epsilon_0}\frac{1}{2R}\left(3 - \frac{r^2}{R^2}\right) & \text{for } r \leq R. \end{cases} \tag{4.1}$$

In the exterior space the potential is identical with that of a point charge; it continues at $r = R$ to a parabolic shape in the interior of the distribution. It is therefore to be expected that in scattering with a sufficiently high energy the scattering cross section would strongly deviate from the Rutherford cross section as soon as the nuclear surface is touched. In addition, the onset of the short-range strong interaction will influence the scattering, especially by absorption. For the calculation of the cross

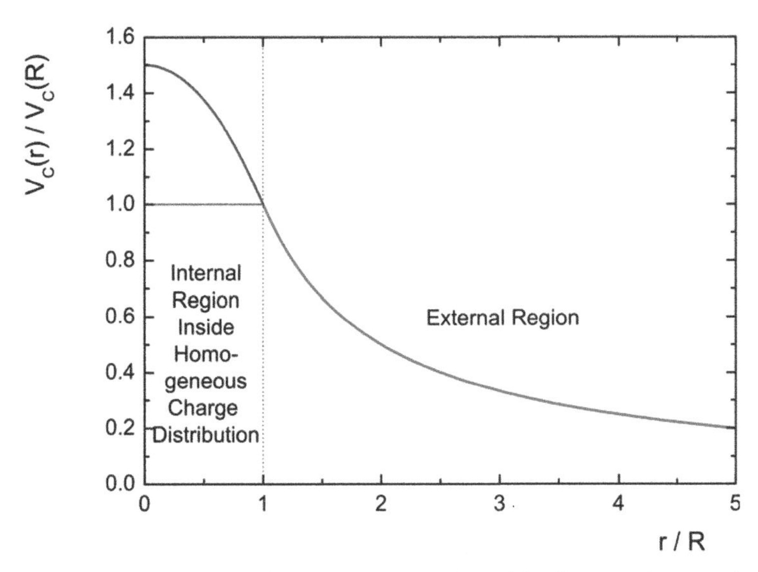

Figure 4.1. Coulomb potential (repulsive for the case of equal signs of the charges such as, e.g., for Rutherford scattering of α or protons from nuclei) of a spherical homogeneous charge distribution. In the external region the potential is that of a point charge at the origin.

section an integral over the contributions from all charge elements $\mathrm{d}q = Ze\rho(\vec{r})\mathrm{d}\tau$ to the potential $U(\vec{r}) = -\frac{Z_1 Z_2 e^2}{R} \cdot \mathrm{e}^{-\alpha R}\rho(\vec{r})\mathrm{d}\tau$ has to be performed:

$$U\left(\vec{r}\,'\right) = -Z_1 Z_2 e^2 \int \rho\left(\vec{r}\,'\right) \frac{\mathrm{e}^{-\alpha R}}{R} \mathrm{d}\tau. \tag{4.2}$$

By inserting this into the Born approximation 2.33 (with $\mathrm{d}\vec{R} = \mathrm{d}\vec{r}\,'$ and $\vec{R} = \vec{r}\,' - \vec{r}$) one obtains:

$$\frac{\mathrm{d}\sigma}{\mathrm{d}\Omega} = \left(\frac{Z_1 Z_2 e^2 m}{2\pi\hbar^2}\right)^2 \cdot \left[\int \rho\left(\vec{r}\right) \mathrm{e}^{\mathrm{i}\vec{K}\vec{r}} \mathrm{d}\tau \cdot \int \frac{\mathrm{e}^{-\alpha R}}{R} \mathrm{e}^{\mathrm{i}\vec{K}\vec{R}} \mathrm{d}\vec{R}\right]^2. \tag{4.3}$$

The cross section factorizes into two parts, one of which (after a transition to the limit $\alpha \to 0$) results again in the point cross section, the other in the form factor:

$$\frac{\mathrm{d}\sigma}{\mathrm{d}\Omega} = \left(\frac{\mathrm{d}\sigma}{\mathrm{d}\Omega}\right)_{\text{point nucleus}} \cdot \left|F\left(\vec{K}^2\right)\right|^2. \tag{4.4}$$

This separation is characteristic for the interaction between extended objects and signifies a separation between the interaction (e.g. the Coulomb interaction) and the structure of the interacting particles.

For rotationally symmetric problems the form factor has a simplified interpretation:

$$F(K) = \int \rho(r) \exp\left(\mathrm{i}\vec{K}\vec{r}\right) 2\pi r^2 \mathrm{d}r \sin\theta \, \mathrm{d}\theta. \tag{4.5}$$

On substitution $u = \mathrm{i}Kr\cos\theta$ and $\mathrm{d}u = -\mathrm{i}Kr\sin\theta\,\mathrm{d}\theta$ this becomes

$$F(K) = 2\pi \int \rho(r)\mathrm{e}^{u} r^2 \mathrm{d}r \frac{\mathrm{d}u}{-\mathrm{i}Kr}$$
$$= \int \rho(r) 4\pi r^2 \mathrm{d}r \cdot \underbrace{\left[\frac{\sin(Kr)}{Kr}\right]}_{\text{purely real}}. \tag{4.6}$$

Thus the form factor is a folding integral of the density with the sampling function (in parentheses). This function is oscillatory and its oscillation 'wavelength' $1/K$ (which depends on the energy of the transferred radiation) has to be adjusted to the rate of change of the density. If the oscillation is too frequent the integral results in ≈ 0 revealing no information on ρ. If it is too slow the sampling function is $\approx$ constant and the integral results in just the total charge Ze. Figure 4.2 illustrates this for different momentum transfers on a given nuclear density distribution. Experimentally the form factor is obtained as the ratio

$$\left(\frac{\mathrm{d}\sigma}{\mathrm{d}\Omega}\right)_{\text{experimental}} \Bigg/ \left(\frac{\mathrm{d}\sigma}{\mathrm{d}\Omega}\right)_{\text{point,theor}}. \tag{4.7}$$

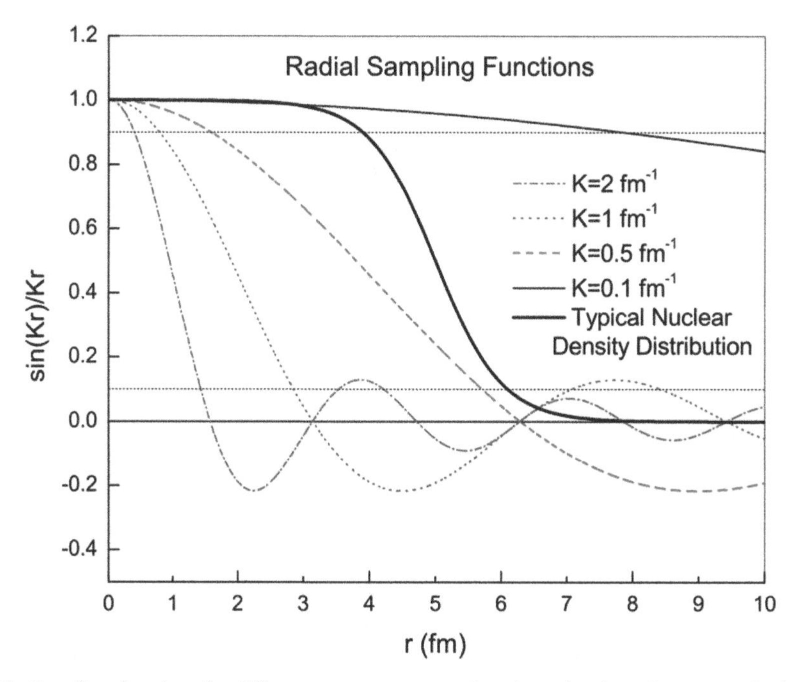

Figure 4.2. Sampling functions for different momentum transfers show that in order to sample details of a given structure (e.g. the shape around the radius of a nuclear density (charge or mass) distribution) the momentum transfer (given by the incident energy and the scattering angle) has to have an appropriate intermediate value. In the example shown the value of $K = 0.5\ \mathrm{fm}^{-1}$ is suitable for sampling the region around the nuclear radius of 5.0 fm. The horizontal dashed lines indicate a 10–90% sampling region.

The charge distribution (or more generally, the density distribution, e.g., of the hadronic matter) is obtained by Fourier inversion of the form factor F:

$$\rho_c(\vec{r}) = \frac{1}{(2\pi)^3} \int_{0\to\infty} F_c\left(\vec{K}^2\right) e^{-i\vec{K}\vec{r}}\, d\vec{K}. \tag{4.8}$$

This means that (in principle) for a complete knowledge of $\rho(\vec{r})$, F must be known for all values of the momentum transfer. Since $\rho(\vec{r})$ for small $\vec{r}$ is governed by the high-momentum transfer components of $\vec{K}$ this cannot be achieved in practice. For this reason the following approximations may be used:

- Model assumptions are made for the form of the distribution: e.g. homogeneously charged sphere, exponential, Yukawa, or Woods–Saxon (also Fermi) behavior.
- The model-independent method of the expansion of $e^{i\vec{K}\vec{r}}$ into moments.

Ansatz for models

It is useful to obtain an impression of the Fourier transformation of different model density distributions as shown in figure 4.3. It is a general observation that 'sharp-edged' distributions lead to oscillating form factors (and therefore cross sections) and smooth distributions to smooth form factors. In agreement with our ansatz a δ distribution (characteristic for a point charge or mass) corresponds to a constant form factor ('scale invariance').

Expansion into moments

With the power-series expansion of $e^{i\vec{K}\vec{r}}$ the form factor becomes

$$F\left(\vec{K}^2\right) \propto \int \rho\left(\vec{r}\right)\left[1 + i\vec{K}\vec{r} - \frac{\left(\vec{K}\vec{r}\right)^2}{2!} \pm \cdots\right] d\tau. \tag{4.9}$$

By assuming a spherically symmetric distribution (with pure r dependence only) and with a normalization such that for a point object the constant form factor is 1, we have:

$$F\left(\vec{K}^2\right) = 1 - \text{const} \cdot K^2 \int_{0\to\infty} r^2 \rho(r)\, d\tau \pm \cdots \tag{4.10}$$

The second term contains the average square radius $\langle r^2 \rangle = r^2_{\text{rms}}$. For small values of $K^2 \langle r^2 \rangle$ one obtains in a model-independent way (i.e. for arbitrary form factors):

$$F\left(\vec{K}^2\right) \approx 1 - \frac{1}{6} K^2 \langle r^2 \rangle. \tag{4.11}$$

Of course this approximation becomes worse with smaller r (because one needs higher moments), i.e. if one wants to resolve finer structures.

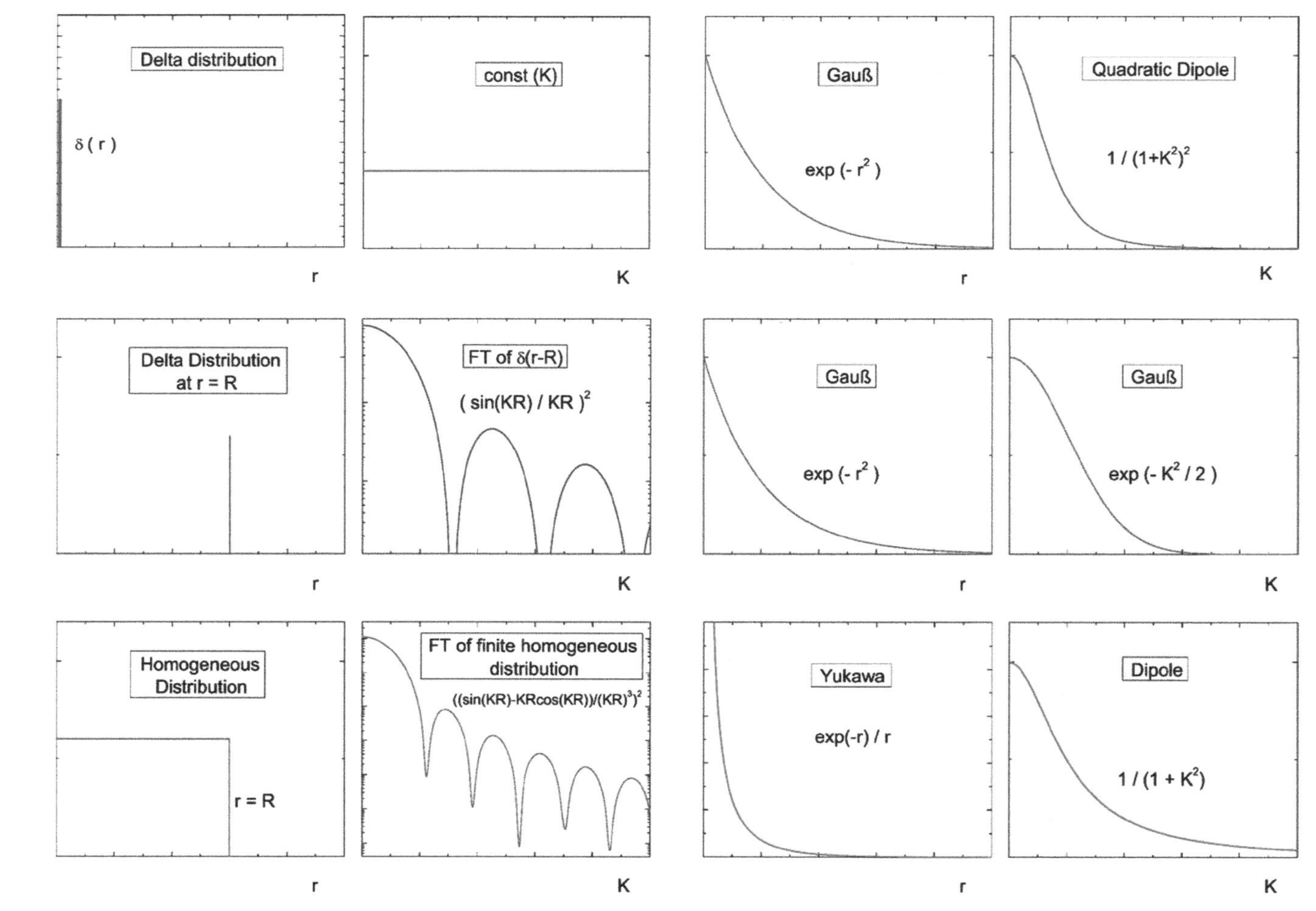

Figure 4.3. Squares of the Fourier transforms—basically the form factors determining the shapes of the differential cross sections—of different model charge density distributions.

4.1 Hadron scattering experiments

After accelerators became available the charge and matter density distributions of nuclei and their radii could be investigated by probing the distributions with hadronic projectiles. With light as well as with heavy ions and also with neutrons as projectiles it is evident that they are extended and possess structure. The consequence is that detailed statements about the density distributions are difficult to make and may need the deconvolution of the contributions from the projectile and target nuclei. However, statements about nuclear radii are possible, even with quite simple semi-classical assumptions such as absorption between nuclei setting in sharply at a well-defined distance and pure Coulomb scattering beyond that distance. Systematic α scattering studies on many nuclei (where we already have strong absorption at the nuclear surfaces) revealed good $A^{1/3}$ systematics for the nuclear radii. A dependence of $\sigma_{\alpha,\alpha}$ on $R = R_0(A^{1/3} + 4^{1/3})$ was fitted to the data, assuming a sharp-cutoff model for the cross sections and taking into account the finite radii of both nuclei. It yielded a radius constant of

$$R_0 = 1.414\,\text{fm}. \tag{4.12}$$

However, when considering the range of the nuclear force for both nuclei of about 1.4 fm a radius constant of $\approx$1.2 fm resulted.

4.1.1 Nuclear radii from higher-energy α-particle scattering

Even without detailed knowledge of the density distribution or of the potential some quite precise statements about nuclear radii by scattering of charged projectiles from nuclei are possible. One condition for this is, however, that the potential, which is responsible for the deviations from the point cross section is of short range, i.e. the charge distribution has a relatively sharp edge.

Most impressively these deviations from the point cross section appear with diminishing distances between the projectile and target in a suitable plot. Because the Rutherford cross section itself is strongly energy and angle dependent one may choose to plot the ratio

$$\left(\frac{\mathrm{d}\sigma}{\mathrm{d}\Omega}\right)_{\text{exp}} \Bigg/ \left(\frac{\mathrm{d}\sigma}{\mathrm{d}\Omega}\right)_{\text{point,theor}} \tag{4.13}$$

as function of the minimum scattering (the *apsidal*) distance d. Thus data at very different energies and angles can be directly compared. If, in addition, one wants to check on the assumption of the systematics of nuclear radii to follow $r = r_0 A^{1/3}$, a universal plot for all possible scattering partners by plotting the above ratio against $d/(A_1^{1/3} + A_2^{1/3})$ is useful. The experimental results show the extension of the charge distribution and the rather sudden onset of (hadronic) absorption (provided the interaction has a strong absorption term, which is typical for $A \geqslant 4$).

Around 1954 α-particle beams from cyclotrons with energies much higher than those from radioactive sources became available, typically from about 20–40 MeV. In the classical Rutherford picture these energies were high enough that the colliding nuclei could be brought into contact in order to 'feel' the (hadronic) nuclear interaction in addition to the Coulomb field. According to this classical model the point of 'grazing' depends on the energy and the scattering angle, see (2.14).

The key experiments were performed at the Brookhaven cyclotron with 40 MeV α beams on heavy nuclei. Using α particles on heavy targets has significant advantages, as compared, e.g., to protons:

- The Coulomb interaction is quite strong.
- Quantum-mechanical interference and diffraction effects are small, often allowing a semi-classical description.
- The onset of strong absorption by the nuclear force is quite abrupt.

Figures 4.4 and 4.5 are from the key paper by Wegner *et al* [30].

The point of deviation from the Rutherford cross section is clearly visible and corresponds to

$$d_{\min} \approx 1.7\,\mathrm{fm}. \tag{4.14}$$

The best description of the data was achieved with the assumption that nuclear radii (including that of the α particle) follow a

$$r = r_0 A^{1/3} \tag{4.15}$$

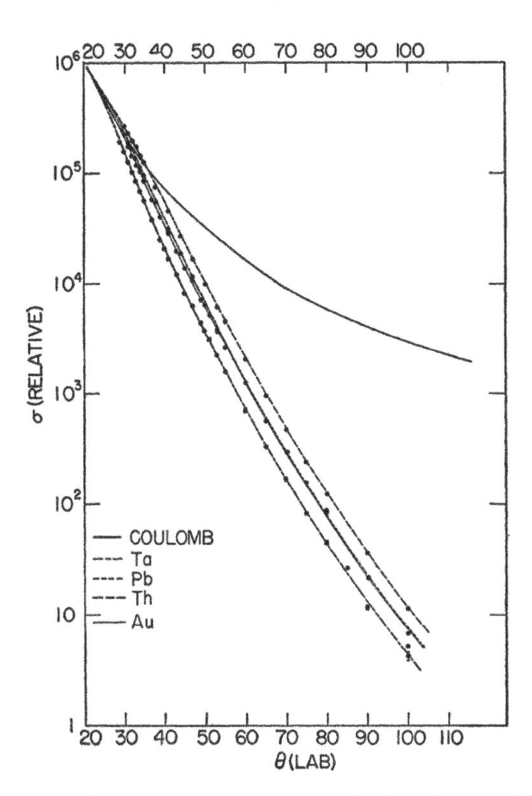

Figure 4.4. Cross-section angular distributions of 40 MeV α scattering from heavy nuclei. The solid line is the pure point-Rutherford prediction. Reproduced with permission from [30]. Copyright 1955 American Physical Society.

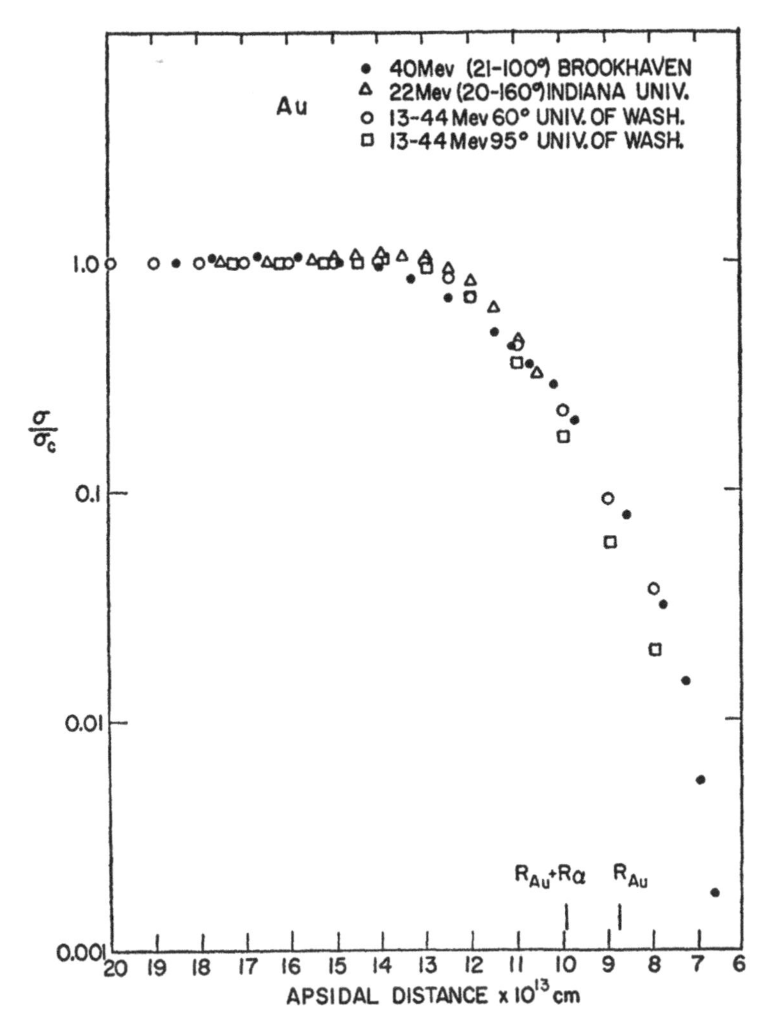

Figure 4.5. The same as figure 4.4, but plotted is the ratio of the measured cross section and the calculated Rutherford cross section against the calculated distance of closest approach ('apsidal distance'), see (2.14), for α particles on Au. This plot shows the relatively sudden onset of absorption by the nuclear interaction together with the effect of the extended charge distribution, which allows, using the $A^{1/3}$ law, the derivation of the range of the nuclear force of about 1.4 fm. The 40 MeV data from Brookhaven [30] have been augmented by 22 MeV data from Indiana University [13] and 13–44 MeV data from the University of Washington [28]. Reproduced with permission from [30]. Copyright 1955 American Physical Society.

law. With the additional assumption of a range of the nuclear force of $\approx$1.4 fm a 'strong' radius constant of

$$r_0 \approx 1.45\,\text{fm} \tag{4.16}$$

resulted.

4.1.2 Heavy-ion scattering

As in α-particle scattering the strong absorption properties of the nuclear interaction in heavy-ion scattering experiments have been very useful to gain insights into nuclear radii and other surface properties. A great number of different pairs of collision partners yielded very good systematics as shown in figure 4.6. This becomes evident especially by plotting the relative cross sections against the distance parameter d, for which an assumed $A^{1/3}$ dependence of the radii of both collision partners was applied

$$d = D_0\left(A_1^{1/3} + A_2^{1/3}\right)^{-1} \tag{4.17}$$

with D_0 the distance of closest approach, as calculated from energies and scattering angles. A well-defined sharp distance parameter of $d_0 = 1.49$ fm for the onset of absorption results. This corresponds to a universal radius parameter of $r_0 = 1.1$ fm if the range of the nuclear force is set to 1.5 fm. The simple model applied was to assume

- Pure point-Rutherford scattering outside the range of nuclear forces.
- A ratio of elastic to Rutherford cross section

$$\frac{\mathrm{d}\sigma}{\mathrm{d}\sigma_R} = 1 + P_{\mathrm{abs}}(D) \tag{4.18}$$

and

$$P_{\mathrm{abs}}(D) = \begin{cases} 0 & \text{for } D \geqslant D_0, \\ 1 - \exp\left(\dfrac{D - D_0}{\Delta}\right) & \text{for } D < D_0, \end{cases} \tag{4.19}$$

with $P_{\mathrm{abs}}(D)$ the probability of absorption out of the elastic channel, D_0 the interaction distance and Δ the 'thickness' of the transition region.

The latter depends on the mass number A of the nuclei involved and could be determined with good accuracy to be, e.g., $\Delta \approx 0.33$ fm for scattering of nuclei near ^{40}Ca from ^{208}Pb. Figure 4.6 shows a large number of heavy-ion combinations in scattering experiments from several authors and a clearcut distance of closest approach where strong absorption sets in.

4.2 Elastic electron scattering—Hofstadter's experiments

Since all electrons (and all leptons) are considered to be point particles they are the ideal projectiles. They 'see' the electromagnetic (and weak) structure of the nuclei. Of course, the treatment must be relativistic. Instead of the Rutherford (point-Coulomb) approach one has to use the proper theory.

In addition to the relativistic treatment, differences to the (classical) Rutherford cross section arise from lepton spin. The derivation of the correct scattering cross

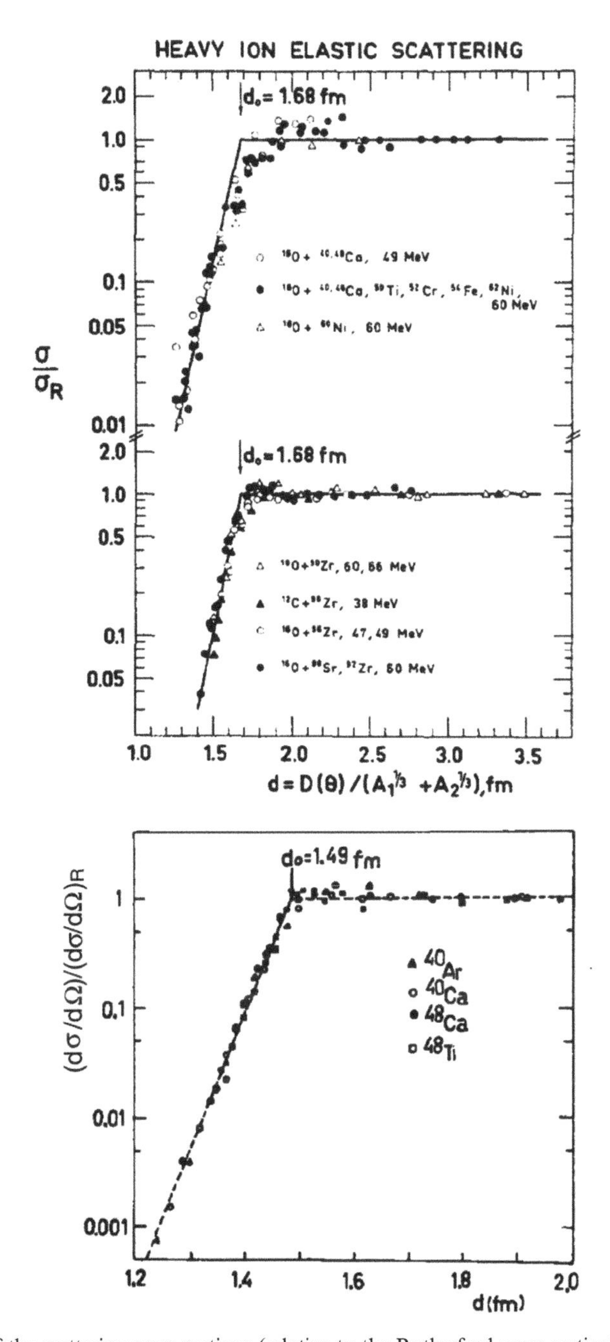

Figure 4.6. Plot of the scattering cross sections (relative to the Rutherford cross section) as functions of the distance of closest approach of many different heavy-ion pairings. Reproduced with permission from [21] (bottom) and [8] (top two). Copyright 1973 and 1978 Elsevier.

section relies on the methods of quantum electrodynamics (QED) and techniques such as the Feynman diagrams. Here only the results will be presented. The electromagnetic interaction between the electron and a hadron is mediated by the exchange of virtual photons, which is accompanied by a transfer of energy and momentum. The wavelength of these photons derives directly from the momentum transfer $\hbar K = 2(h\nu/c)\sin(\theta/2)$ to be

$$\lambdabar_{\text{de Broglie}} = \frac{\hbar}{\hbar K} = 1/K. \tag{4.20}$$

The argument of diffraction limitation may also be formulated in the complementary time picture; it may be said that at long wavelengths, due to the uncertainty relation, one needs long measurement times in which the projectile sees only a time-averaged picture of the object considered while small wavelengths allow measurement times equivalent to snapshots of the object or its substructures (partons).

Principally in lepton scattering at higher energies three distinct regions of momentum transfer can be distinguished:

- Elastic scattering at small momentum transfer is suitable for probing the shape of the hadrons. The resulting two form factors produce again the charge and current (magnetic moment) distributions and the radii of the hadrons by Fourier inversion.
- (Weakly) inelastic (also called quasi-elastic) scattering at higher momentum transfer leads to excitations of the hadrons (e.g. Delta or Roper excitations (resonances) of the nucleons). The form factors are quite similar to those from elastic scattering, which means that we have some excited state of the same nucleons. A key experiment showing the excitation spectrum of the proton by inelastic electron scattering at DESY-Hamburg is that of [2], as displayed in figure 4.7.
- Deep-inelastic scattering is the method suited for seeing partons inside the hadrons. Using this method of electron and muon scattering the quarks bound in nucleons and their properties (spin, momentum fraction) and also the existence of sea quarks (s-quark–anti-quark pairs) were identified. In particular, the point-like character of these constituents was shown by the near constancy of the form factors (here called *structure functions* with the momentum transfer (Bjorken scaling [5])), see also chapter 6.

Here only elastic scattering will be discussed in detail. In QED theory for the differential cross section the Rosenbluth formula was deduced:

$$\frac{d\sigma}{d\Omega} = \left(\frac{d\sigma}{d\Omega}\right)_{\text{point}} \cdot \left(\frac{F_E^2 + bF_M^2}{1+b} + 2bF_M^2 \tan^2\frac{\theta}{2}\right). \tag{4.21}$$

The point cross section $(d\sigma/d\Omega)_{\text{point}}$ is a generalized Rutherford cross section and is calculable with the methods of QED (e.g. using Feynman diagrams). The most general form of this cross section (the Dirac scattering cross section) contains as its main part the electrostatic scattering, a contribution from the magnetic

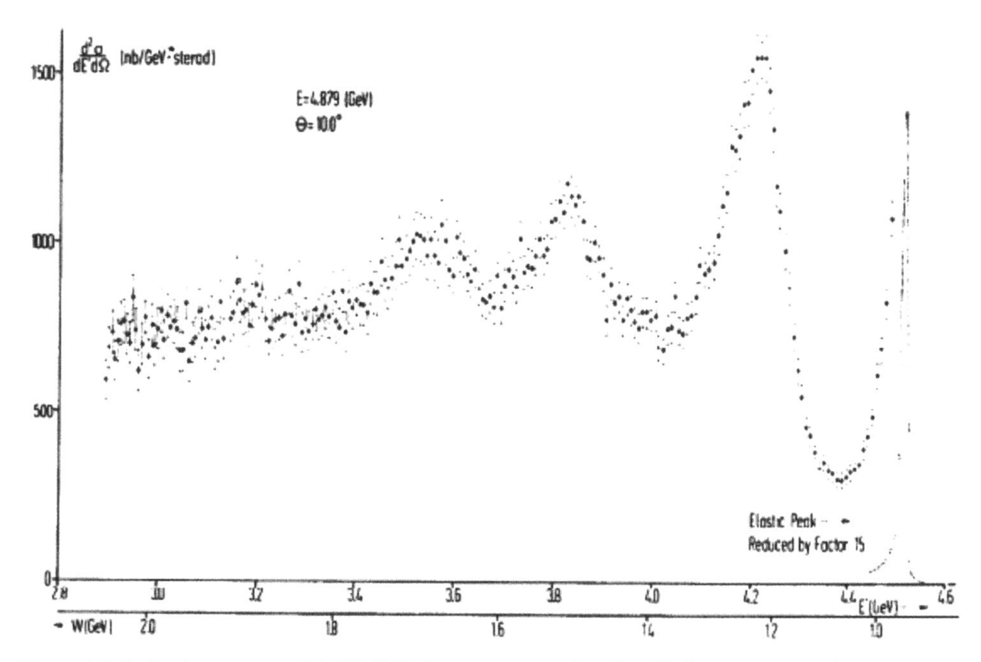

Figure 4.7. Inelastic spectrum of 4.879 GeV electrons scattered at $\theta = 10°$ from protons at low momentum transfer at the DESY facility in Hamburg. Starting from the right there is the (15 times reduced) elastic peak, then pion–nucleon resonances with the recoiling-state masses of 1.24 GeV, the Δ resonance, followed by the 1.51 and 1.69 GeV resonances. Reproduced with permission from [2]. Copyright 1968 Elsevier.

(spin-dependent) interaction, which depends on the momentum transfer, and a correction for the nuclear recoil:

$$\left(\frac{d\sigma}{d\Omega}\right)_{\text{Dirac}} = \frac{\alpha^2}{4p_0^2 \sin^4(\theta/2)}\left(1 + \frac{2p_0}{M}\sin^2\frac{\theta}{2}\right)\left(\cos^2\frac{\theta}{2} + \frac{q^2}{2M^2}\sin^2\frac{\theta}{2}\right). \tag{4.22}$$

For small energies or momentum transfers the cross section simplifies to:

$$\left(\frac{d\sigma}{d\Omega}\right)_{\text{Mott}} = \frac{\left[2e^2\left(E'c^2\right)\right]^2}{q^4} \cdot \frac{E'}{E}\cos^2\frac{\theta}{2}, \tag{4.23}$$

where q = four-momentum transfer, $b = -q^2/(4\,m^2c^2)$, and E' and E are the energies of the outgoing and incoming electrons. F_E and F_M are the electric and magnetic form factors of the nucleons. Experimentally they are obtained from the measured data by least-squares fitting of the parameters of the theory, graphically through the Rosenbluth plot, i.e. by plotting $(d\sigma/d\Omega)_{\text{exp}}/(d\sigma/d\Omega)_{\text{point}}$ against $\tan^2(\theta/2)$.

In analogy to the Rutherford cross section, here the form factors (or structure functions) are Fourier transforms of the charge and current density distributions (or distributions of the (anomalous) magnetic moments). As previously, these distributions result from Fourier inversion of the form factors and at the same time quantitative values of the shapes and sizes of the nucleons are obtained.

The measured form factors as functions of q^2 are normalized such that for $q \to 0$ they become the static values of the electric charge and magnetic moments. Except for the electric form factor of the neutron all others are well described by the dipole ansatz corresponding to a density distribution of an exponential function.

An early model for the charge density distribution was—in addition to the homogeneously charged sphere with only one parameter, its radius—a modified Woods–Saxon distribution with three parameters, because, in addition to the radius parameter r_0 and the surface thickness a, the central density ρ_0 must also be adjustable because it varies, in particular in light nuclei:

$$\rho_c(r) = \frac{\rho_0}{1 + e^{\frac{r - r_{1/2}}{a}}}. \tag{4.24}$$

The surface thickness $t = 4 \ln 3 \cdot a$ signifies the 10–90% thickness range centered around $r_{1/2}$. From this parametrization an electromagnetic radius constant of $r_{1/2} = 1.07\,\text{fm}$, a surface-thickness parameter of $a = 0.545\,\text{fm}$ and a central density of $\rho_N = 0.17$ nucleons fm^{-3} or $1.4 \cdot 10^{14}\,\text{gcm}^{-3}$ for nuclei with $A > 30$ have been derived. The description of 'modern' density distributions is not so simple because the nuclei have an individual and more complex structure even if the essential features such as the three parameters do not vary too much. Detailed structure information is obtained from model-independent approaches such as Fourier–Bessel expansions. Radii are given as rms radii or are converted into the equivalent radii R_0. R_0 is the radius of a homogeneously charged sphere of equal charge using the relation

$$r_{\text{rms}} = \sqrt{3/5}\, R_0. \tag{4.25}$$

The definition of the (model-independent) Coulomb rms radius is

$$r_{\text{rms}} = \left\langle r^2 \right\rangle^{1/2} = \left(\frac{1}{Ze} \int_0^\infty r^2 \rho_C(r) 4\pi r^2 \text{d}r \right)^{1/2}. \tag{4.26}$$

The electron scattering experiments by Douglas Hofstadter and co-workers at Stanford were key experiments for the elucidation of the extended nature of nucleons and nuclei, their charge and mass distributions, radius systematics, surface thickness, range of nuclear forces, etc. Later these experiments led to the experimental verification of the point-like structures inside nucleons, the quarks, and their properties. The methods used included linear electron accelerators, among them Marc III with an energy of 1 GeV, then the powerful two-mile-long SLAC accelerator with an energy of 20 GeV equipped with large spectrometers, as depicted in figures 4.8 and 4.9. Some earlier results are shown in figures 4.10 and 4.11. For the study of nucleons higher-energy electrons were used. Figure 4.11 depicts the differential elastic scattering cross section with fits using electric and magnetic form factors in an exponential model [6] with point and extended nucleon assumptions. The results of elastic electron scattering on the proton show that the proton—differing from heavier

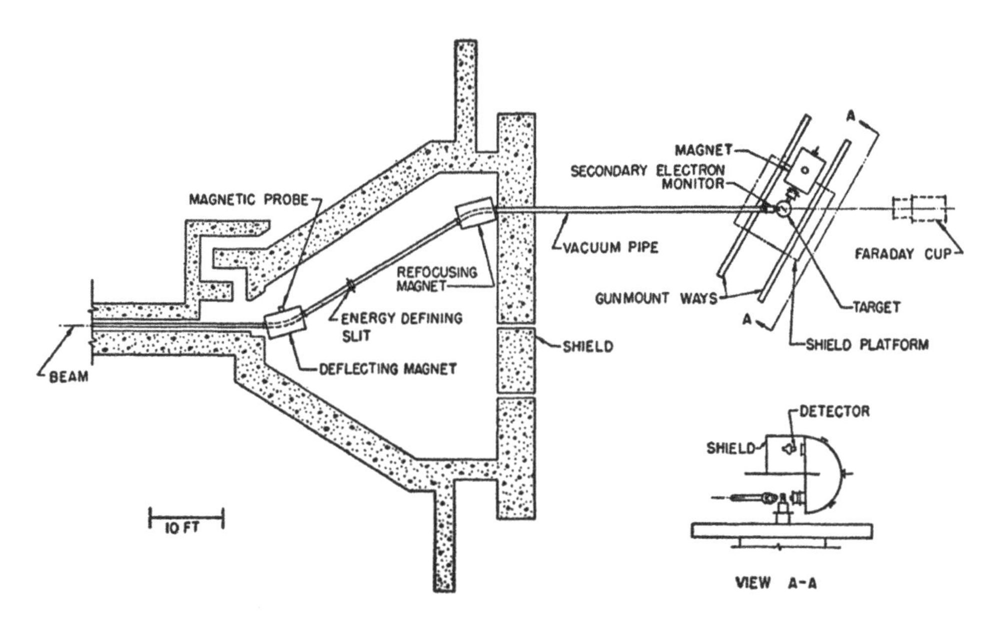

Figure 4.8. The Stanford facility. Reproduced with permission from [6]. Copyright 1956 American Physical Society.

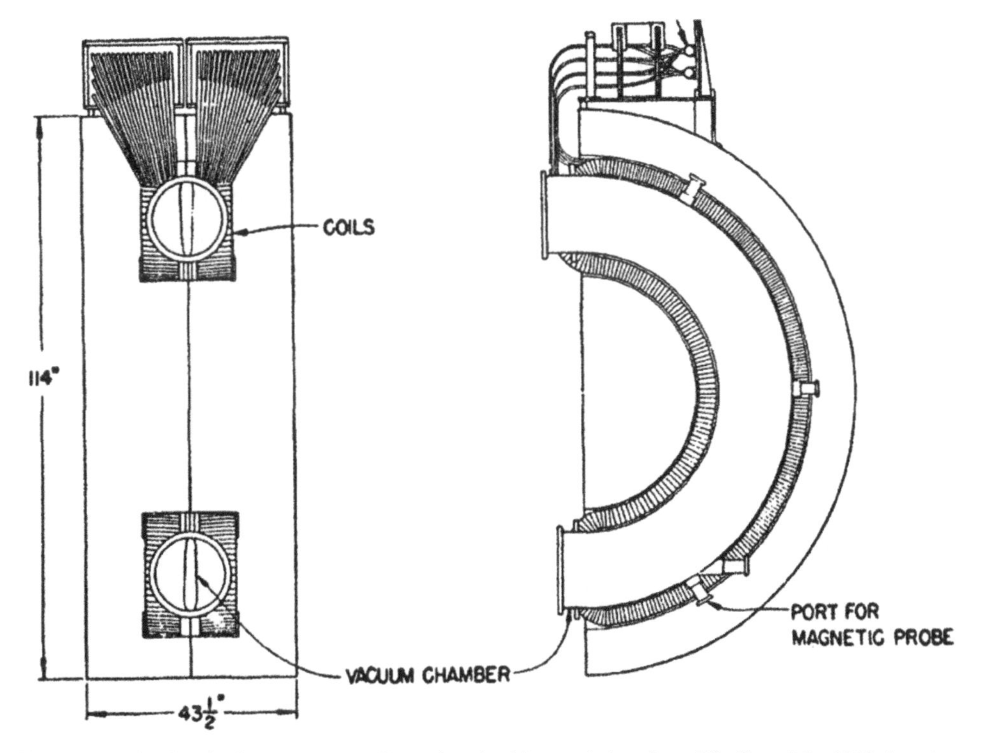

Figure 4.9. The Stanford spectrometer. Reproduced with permission from [6]. Copyright 1956 American Physical Society.

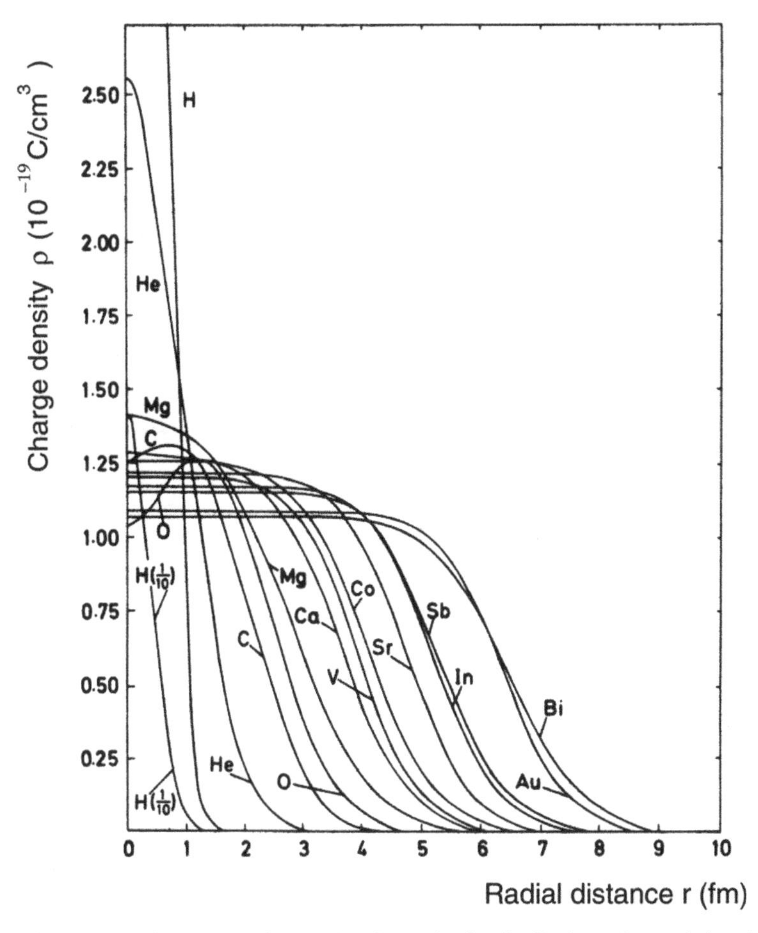

Figure 4.10. Electron scattering cross-section results: charge density distributions of extended nuclei across the periodic table. The cross-section formula is the simplified Mott cross-section formula 4.23, amended by a form-factor term. Various model distributions have been applied. For heavier nuclei Fermi distributions describe the densities well. The figure was redrawn for clarity from Hofstadter's Nobel lecture 1961 [18], see also [17].

nuclei—has no sharp surface. Its density distribution has been described appropriately by an exponential with an rms radius of

$$r_{\rm rms} \approx 0.888\,\text{fm}. \tag{4.27}$$

4.3 Key experiments with complementary methods

The scattering methods for determining density distributions and the radii of nuclei are complemented well by methods which rely on the influence of the extended nuclear charge distribution on atomic levels. Laser spectroscopy methods as well as exotic atom methods are sensitive enough to compete with them. As a model the interaction of an extended (homogeneous, spherical) charge distribution with atomic electrons shows the salient features of the modifications from a point charge. Figure 4.12 depicts the charge density distribution together with the potential as

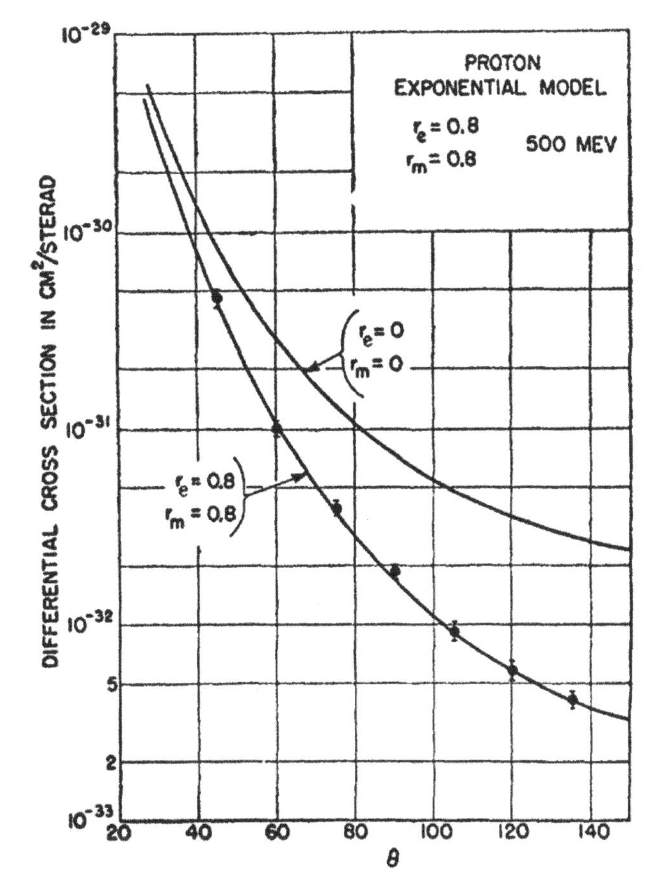

Figure 4.11. Differential cross section of the elastic scattering of 500 MeV electrons on the proton fitted by an exponential form factor. Reproduced with permission from [6]. Copyright 1956 American Physical Society.

functions of r. The Coulomb-energy difference ΔE_C is determined by the integral over the potential difference for the two cases

$$\Delta E_C = \int_0^R e\rho_e(r)\Delta\Phi_N(r)4\pi r^2 \mathrm{d}r \tag{4.28}$$

$$\cong 4\pi e \underbrace{\rho_e(r)}_{|\psi_e(0)|^2} \int_0^R \left[\underbrace{\frac{C}{2R}\left(3 - \frac{r^2}{R^2}\right)}_{\text{homogeneous}} - \underbrace{\frac{C}{R}}_{\text{point}} \right] r^2 \mathrm{d}r \tag{4.29}$$

$$\approx -\left|\psi_e(0)\right|^2 \cdot \frac{Ze^2}{4\pi\epsilon_0} \cdot \frac{4\pi R^2}{10}. \tag{4.30}$$

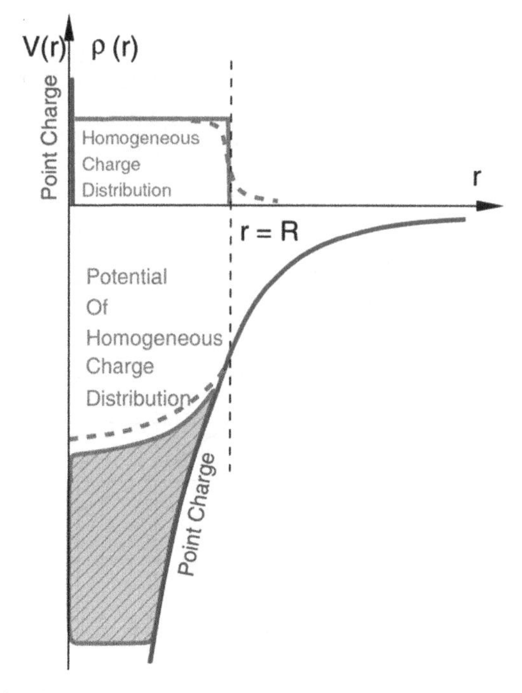

Figure 4.12. Charge density distribution and Coulomb potential (attractive for charges of opposite signs) of a point charge compared to an extended homogeneous charge distribution (with sharp and 'soft' edges) and its potential.

For the 1S ground state of the hydrogen atom the electron's radial wave function is

$$\psi^{n\ell m} \propto \left(\frac{Z}{a_0}\right)^{3/2} \tag{4.31}$$

leading to

$$\Delta E_C \approx -\frac{2}{5} \cdot \frac{Z^4}{4\pi\epsilon_0} \cdot e^2 \cdot \frac{R^2}{a_0^3}, \tag{4.32}$$

assuming a 'contact interaction' at the nuclear center.

4.3.1 High-precision laser spectroscopy

Although the effects of nuclear size and shape on the energies and transitions of atomic electrons are very small the very high precisions reached in laser spectroscopy measurements allow one to achieve results comparable to other methods (such as muonic atoms). The most intense efforts have been focused on the spectroscopy of the hydrogen atom, especially on the measurement of the famous Lamb shift, i.e. the energy separation between the 2S and 2P states, which are predicted to be degenerate in Dirac theory but split by different effects of QED (among them vacuum fluctuations and polarization of the vacuum). This is why the measurement of the Lamb shift to very high precision is essential.

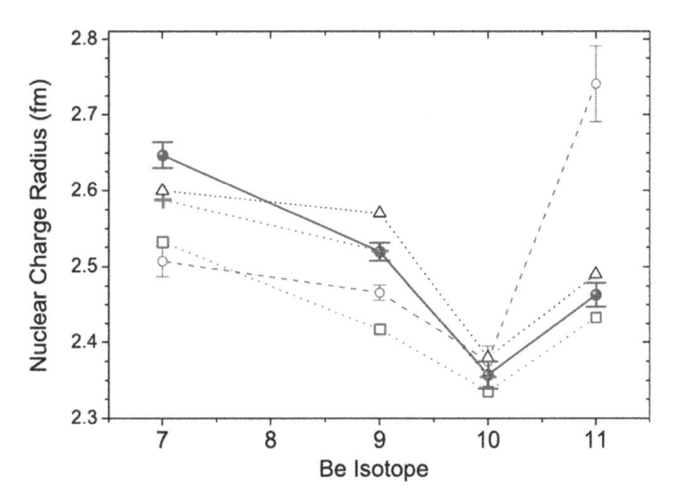

Figure 4.13. Experimental charge radii of Be isotopes: precision laser isotope shift measurements on a Be^+ beam (full dots) compared to interaction cross-section measurements (open circles) and different *ab initio* calculations. Reproduced with permission from [20]. Copyright 2009 American Physical Society.

The interpretation of the measurements requires the evaluation of the influence of nuclear effects, especially the size (radius and shape) of the nucleus. Currently, the precision of the measurements (in atomic spectroscopy as well as in muonic atoms) is now so high that, e.g., the proton radius has become the final limitation to higher precision of the Rydberg constant and comparisons to QED. As an example for the application of precision laser spectroscopy in connection with the neutron radii of halo nuclei (see chapter 5) we cite [20] in which measurements of the charge radii of ^{7,9,10,11}Be show full agreement with electron scattering and muonic atom results, but expose a significant difference with the neutron (or matter) radii of the halo nuclei. Figure 4.13 shows these results. The results obtained for the proton radius agree within the error margins with those from medium-energy electron scattering. However, a very recent, very puzzling disagreement (of 5 standard deviations!) with results from muonic atoms is unresolved, see the following subsection.

4.3.2 Muonic atoms

The 1S ($n = 1$) *Bohr radius* of a negative particle with mass m a_0 is $\propto m^{-1}$ thus

$$\frac{a_0(\mu^-)}{a_0(e)} = \frac{m_e}{m_{\mu^-}} = 4.75 \cdot 10^{-3} \tag{4.33}$$

and

$$\frac{\Delta E_C(\mu^-)}{\Delta E_C(e)} = \frac{|\psi_\mu(0)|^2}{|\psi_e(0)|^2} = \left[\frac{a_0(e)}{a_0(\mu)}\right]^3 \tag{4.34}$$

$$= \left(\frac{m_\mu}{m_e}\right)^3 = 210.5^3 = 9 \cdot 10^6, \tag{4.35}$$

testifying to a large increase in the sensitivity to the positions of the energy levels and the transitions between them for muonic atoms as compared to electronic ones. However, because the level energies are larger by a factor of $m_\mu/m_e = 210.5$, the transitions are in the x-ray region.

The first experiments with muons (the term 'μ' mesons is erroneous because muons are leptons, mesons are hadrons underlying the strong interaction) could be performed in the 1950s after cyclotrons with sufficient energy became available to produce pions (π mesons), which in turn decay weakly via

$$\pi^- \rightarrow \mu^- + \bar{\nu}_\mu + 34\,\text{MeV} \tag{4.36}$$

(lifetime $\tau = 2.603 \cdot 10^{-8}$ s, corresponding to a flight path of $c\tau = 7.8$ m for highly relativistic pions). By magnetic deflection and time-of-flight techniques the muons are separated from electrons and formed into a muon beam in a muon channel. Their lifetime at rest is $\tau = 2.197 \cdot 10^{-6}$ s and $c\tau = 658.65$ m.

In addition to the properties of the muons, their interactions with nuclei were studied. Negative muons, after slowing down by energy-loss processes in matter, may be captured by the nuclear Coulomb field and forced into outer Bohr orbits of the muonic atoms formed, from where they cascade down into the ground state, thereby emitting characteristic x-rays. The references for the earliest such investigations are [9, 14, 31]. Figures 4.14 and 4.15 show the apparatus used for these experiments. In [14] muonic x-rays on a number of nuclei were measured, of which two examples are shown in figures 4.16 and 4.17. From the shift between the 2p–1s transition energies of the point and extended charge distributions the nuclear radius systematics of $r = r_0 A^{1/3}$ was established with a best value of

$$r_0 = 1.17 \ldots 1.22\,\text{fm} \tag{4.37}$$

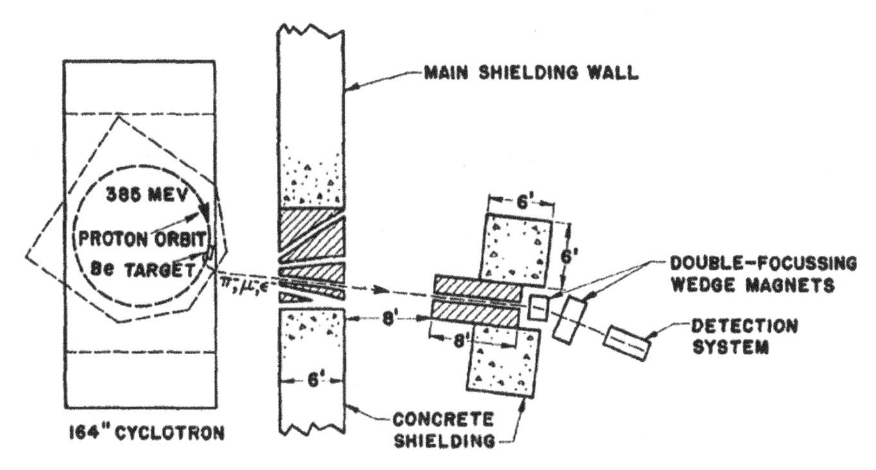

Figure 4.14. Experimental set-up for the production of a beam of muons in Columbia University's 164", 385 MeV cyclotron, separating them from pions and electrons, and slowing down and capturing the μ into Bohr orbits. Reproduced with permission after [14]. Copyright 1953 American Physical Society.

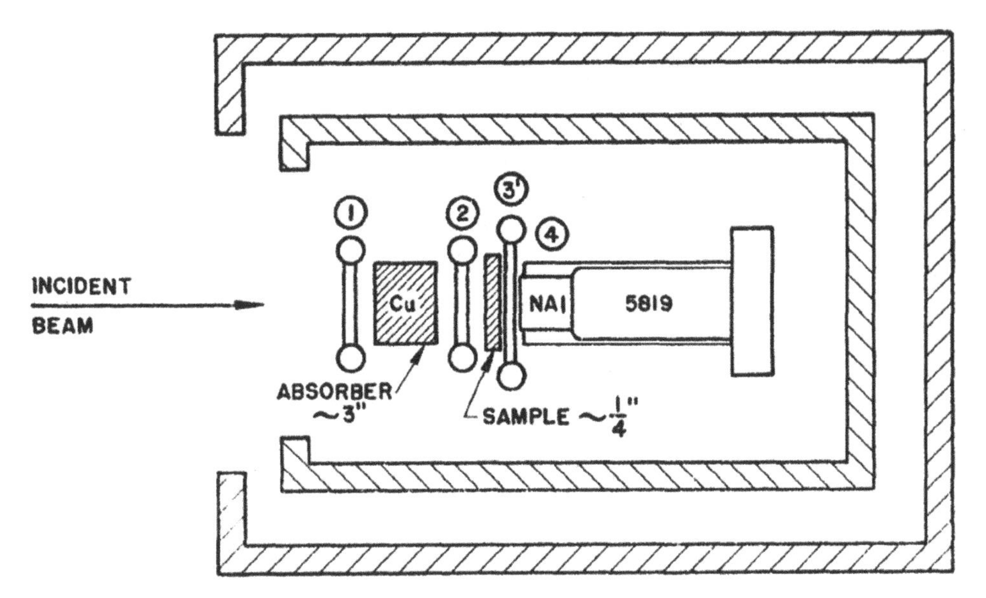

Figure 4.15. Scintillation detector set-up for the measurement of muonic x-rays, emitted in 2p–1s transitions to the ground states of different muonic atoms. Stilbene scintillation counters 1, 2 and 3′ select particles stopped in the sample, the NaJ scintillation detector 4 measures photons and the Cu absorber removes the pions from the beam. The double-walled magnetic shielding consists of 1/2 inch iron plates. Reproduced with permission after [14]. Copyright 1953 American Physical Society.

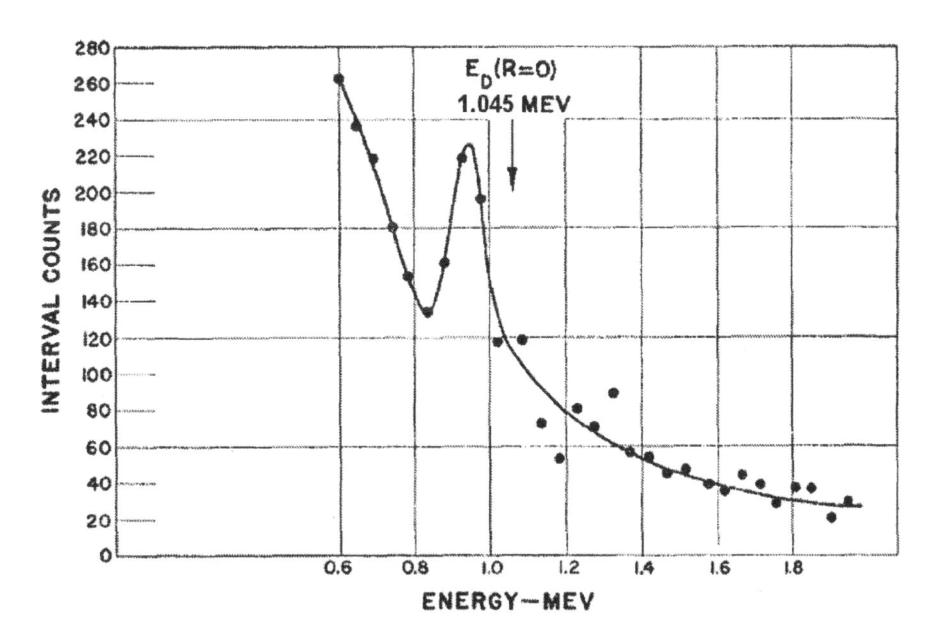

Figure 4.16. Muonic x-ray spectrum of Ti obtained with a NaJ scintillation detector and showing the large energy shift between a point charge and the actual extended charge distributions. Reproduced with permission after [14]. Copyright 1953 American Physical Society.

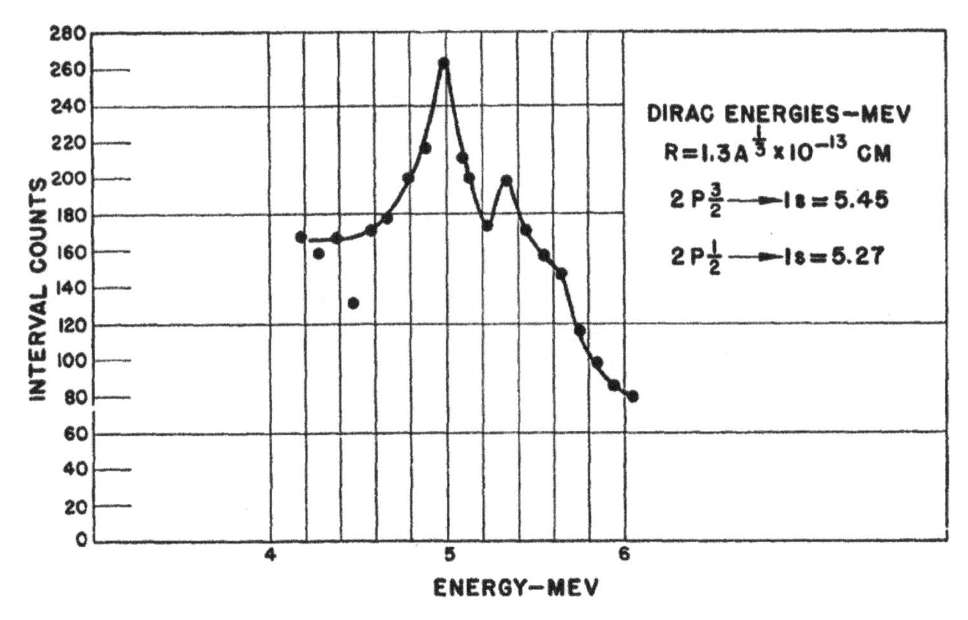

Figure 4.17. Muonic x-ray spectrum of Pb obtained with a NaJ scintillation detector and showing the large energy shift between a point charge and the actual extended charge distributions. The two peaks at 5 MeV and 5.3 MeV are interpreted as pair-production x-ray lines caused by the transition from the $2p_{3/2}$ and $2p_{1/2}$ states to the 1 s_0 state. The two lines are due to the fine-structure splitting expected for the μ being a spin-1/2 lepton of 210 electron masses and are shifted from the point (Dirac) energy by 172% of the measured values. Reproduced with permission after [14]. Copyright 1953 American Physical Society.

and a muon mass of $m_\mu = 210m_e$. More modern methods used Ge(Li), Si(Li), HPGe, crystal spectrometers and LAAPD photodiode detectors.

Nuclear radii from muonic atoms are often more precise than those from lepton scattering but they are in a way complementary in relation to the radius region probed (they measure different moments). Thus, the results of both methods can be combined. An example is shown in figure 4.18. The main features of the distributions are in general quite well reproduced by the early 'mean-field' calculations, but show more structure in the nuclear interior, see also [10, 16]. The salient results of these investigations are:

- From the distributions a central density is derived, which for heavier nuclei is constant in first approximation. This and the systematics of radii are characteristic for nuclear forces. Their properties are: a short range, and saturation and incompressibility of nuclear matter, and suggest an analogy to the behavior of liquids, which led to the development of collective nuclear models (liquid-drop models, and models of nuclear rotation and vibration).
- The radii of spherical nuclei follow more or less a simple law $r = R_0 A^{1/3}$. For the radius parameter $R_0 = 1.24$ fm is a good value. From Coulomb-energy

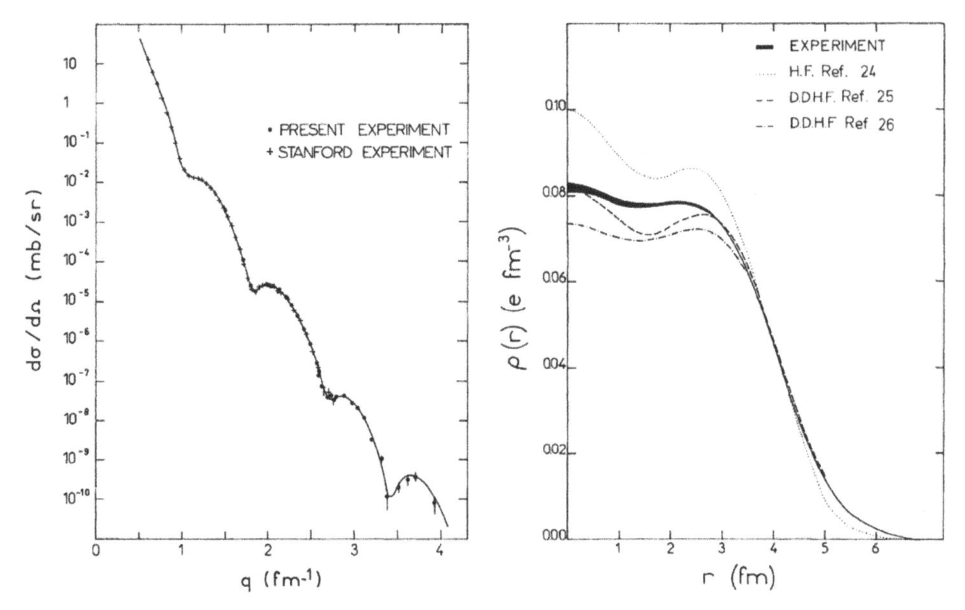

Figure 4.18. Cross section of elastic electron scattering from ^{58}Ni and charge density distribution derived from electron scattering and muonic atom data combined. Reproduced with permission from [26]. Copyright 1975 American Physical Society.

differences of mirror nuclei a value of $R_0 = (1.22 \pm 0.05)$ fm has been derived.

- The surface thickness of all nuclei is nearly constant with a 10–90% value of $t = 2.31$ fm corresponding to $a = t/4 \ln 3 = 0.53$ fm. This is explained by the range of the nuclear forces independent of the nuclear mass number A.
- The nucleons have no nuclear surface. The charge and current as well as the matter densities of the proton follow an essentially exponential distribution. For the neutron the charge distribution is more complicated because the volumes of the negative and positive charges must compensate each other to zero notwithstanding some complicated internal charge distribution that originates from its internal quark–gluon structure.
- The rms radii for the current distributions of protons and neutrons and the charge distribution of the protons are 0.88768(69) fm (accepted CODATA value) in agreement between atomic spectroscopy and electron scattering results. Recently, with increased experimental precision an unresolved discrepancy between values from lepton scattering and muonic atom work has been published [19, 23], see figure 4.19. The rms charge radius of the neutron is -0.1161 ± 0.0022 fm [11], which—with total charge zero—means that there must be positive and negative charges distributed differently over the nuclear volume.
- Thus, nucleons are not 'elementary', but have complicated internal structures. In fact, we now know much more: all hadrons are composed of quark

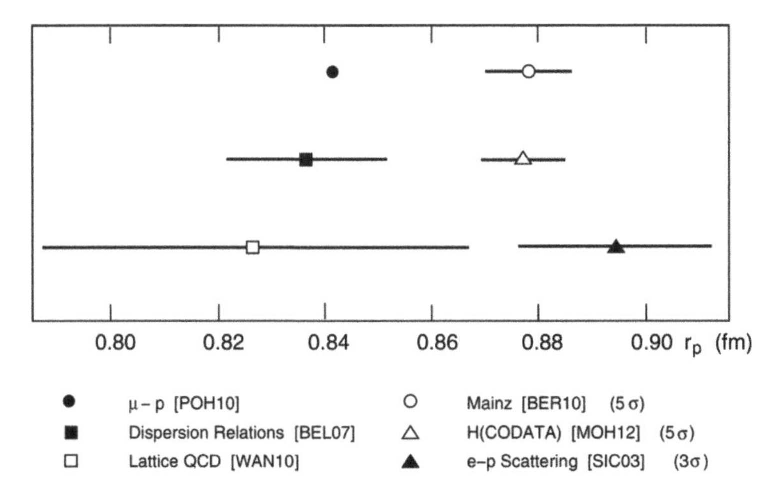

Figure 4.19. Unresolved discrepancies between determinations of the proton's rms radius by different methods. The accepted CODATA-10 value is $r_{\rm rms}(p) = 0.8775(51)$ fm [19] whereas a new muonic atom value is $r_{\rm rms}(p) = 0.84184(67)$ fm [23] while a recent model-independent analysis yields $r_{\rm rms}(p) = 0.8412(15)$ fm [22]. For the dispersion relation result see [3], for the lattice quantum chromodynamics (QCD) calculation see [29], for the recent Mainz measurement see [4] and for e–p scattering see [25].

constituents, three for baryons, two (quark–antiquark pairs) for mesons, bound together by the exchange of gluons. In addition, there is a *sea* of virtual quark–antiquark (s and anti-s) pairs. The Pauli exclusion principle is satisfied by assigning each quark a new quantum number, *color*, and requiring that only color-neutral particles exist in nature, thus free single quarks do not exist (they remain *confined* inside the particles). To 'see' the constituents, high-energy probes are necessary (see the preceding text), but at low or intermediate energies only average quantities (radii, charge, spin, meson-exchange clouds, masses, etc) are visible. The pertinent models for the hadron structure and interactions are therefore meson-exchange models with more recent attempts to base these on effective-field approaches.

4.3.3 Matter density distributions and radii

The matter density and thus the neutron density distributions—apart from and independent of the charge or current distributions—can be investigated only by additional hadronic scattering experiments because neutrons and protons in principle need not have the same distributions in nuclei. A number of medium-energy (e.g. 1 GeV) proton scattering experiments have been performed, see e.g. [7] where charge and neutron radii were compared for a number of nuclei with no significant differences between either.

4.3.4 Hadronic radii from neutron scattering

The total cross sections of 14 MeV neutron scattering under simple assumptions have been shown to also follow a $A^{1/3}$ law, see e.g. [24], p 32, cited from [12].

The assumptions were that the sharp-edged range of the nuclear force was 1.2 fm and the total cross section σ_{tot} follows $2\pi(R + ƛ)^2$ with R the nuclear (hadronic) radius, i.e. the nuclei are considered to be black (totally absorbent) to these neutrons, which is not exactly fulfilled, as the structures in this dependence show. These can be explained with the optical model, see below. The radius constant extracted from this systematics is

$$R_{\text{hadr}} = 1.4\,\text{fm}. \tag{4.38}$$

In addition, there have been attempts to extract the neutron radius of ^{208}Pb from parity-violating electron scattering [1].

4.3.5 Special cases—neutron skin

In particular, the question of a neutron skin in nuclei with neutron excess is interesting and only recently such a thin skin was consistently shown to exist, see e.g. [27] and references therein. Among the hadronic probes used have been protons, α, heavy ions, antiprotons and, recently, also pions, e.g. on ^{208}Pb, ^{40}Ca and others. The extraction of rms radii requires some model assumptions concerning the reaction mechanism and the interplay of hadronic and Coulomb interactions. The pion results are derived from two sources: pionic atoms (in analogy to the derivation of the electromagnetic radii from muonic atoms) and total reaction cross sections of π^+ [15]. The neutron skin is related to the symmetry energy, which plays a role in the Bethe–Weizsäcker mass formula for the binding energies of nuclei, especially for 'asymmetric' nuclei with strong neutron excess, but also for astrophysics and nuclear matter calculations. The radius of neutron stars is closely related to the symmetry energy value in high-density nuclear matter, see e.g. [27].

Usually the quantity

$$\delta R_{\text{np}} = \langle r^2 \rangle_{\text{n}}^{1/2} - \langle r^2 \rangle_{\text{p}}^{1/2} \tag{4.39}$$

is given. The experimental values deduced from different experiments are on the order of $\delta R_{\text{np}} \approx 0.2$ fm.

4.3.6 Neutron versus proton distributions

Whereas the charge and magnetic distributions are best obtained with charged projectiles, for which the interaction is exactly known (e.g. electrons, which do not interact via the strong force), for neutrons one needs nuclear scattering models (e.g. the optical model, see section 10.1). The assumption that the neutron and proton radii of heavier nuclei are about equal has proved too simple with the evidence of neutron halo and neutron skin nuclei, see chapter 5.

Bibliography

[1] Abrahamyan S *et al* (PREX Collaboration) 2012 *Phys. Rev. Lett.* **108** 112502

[2] Bartel W, Dudelzak B, Krehbiel H, McElroy J, Meyer-Berkhout U, Schmidt W, Walther V and Weber G 1968 *Phys. Lett.* B **28** 148

[3] Belushkin M A *et al* 2007 *Phys. Rev.* C **75** 035202
[4] Bernauer J C *et al* 2010 arXiv: 1007.5076
Bernauer J C *et al* 2010 *Phys. Rev. Lett.* **105** 242001
[5] Bjorken J D 1967 *Phys. Rev.* **163** 1767
[6] Chambers E E and Hofstadter R 1956 *Phys. Rev.* **103** 1454
[7] Chaumeaux A, Layly V and Schaeffer R 1977 *Phys. Lett.* B **72** 33
[8] Christensen P R, Manko V I, Becchetti F D and Nickles R J 1973 *Nucl. Phys.* A **207** 33
[9] Cooper L and Henley E 1953 *Phys. Rev.* **92** 801
[10] Dechargé J and Gogny D 1968 *Phys. Rev.* C **21** 1568
[11] Eidelman S *et al* (Particle Data Group) 2004 *Phys. Lett.* B **592** 1
[12] England J B A 1974 *Techniques in Nuclear Structure Physics* (New York: Halstead)
[13] Farwell G W and Wegner H E 1954 *Phys. Rev.* **93** 356
Farwell G W and Wegner H E 1954 *Phys. Rev.* **95** 1212
[14] Fitch V L and Rainwater J 1953 *Phys. Rev.* **92** 789
[15] Friedman E 2012 *Nucl. Phys.* A **896** 46
[16] Frois B and Papanicolas C N 1987 *Ann. Rev. Nucl. Part. Sci.* **37** 133
[17] Hofstadter R 1957 *Ann. Rev. Nucl. Sci.* **7** 231
[18] Hofstadter R 1961 *Nobel Lecture*, the Nobel Foundation (Stockholm) and (Amserdam: Elsevier)
[19] Mohr P J, Taylor B N and Newell D B (CODATA-10) 2012 *Rev. Mod. Phys.* **84** 1527
[20] Nörtershäuser W *et al* 2009 *Phys. Rev. Lett.* **102** 062503
[21] Oganessian Yu Ts, Penionzhkevich Yu E, Man'ko V I and Polyanski V N 1978 *Nucl. Phys.* A **303** 259
[22] Peset C and Pineda A 2015 *Eur. Phys. J.* A **51** 32
[23] Pohl R 2010 *Nature* **466** 213
[24] Satchler G R 1990 *Introduction to Nuclear Reactions* 2nd edn (London: McMillan)
[25] Sick I 2003 *Phys. Lett.* B **576** 62
[26] Sick I, Bellicard J B, Bernheim M, Frois B, Huet M, Leconte Ph, Mougey J, Xuan-Ho P, Royer D and Turck S 1975 *Phys. Rev. Lett.* **35** 910
[27] Tsang M B *et al* 2012 *Phys. Rev.* C **86** 015803
[28] Wall N S, Rees J R and Ford K W 1955 *Phys. Rev.* **97** 726
[29] Wang P *et al* 2009 *Phys. Rev.* D **79** 094001
[30] Wegner H E, Eisberg R M and Igo G 1955 *Phys. Rev.* **99** 825
[31] Wheeler J A 1953 *Phys. Rev.* **92** 812

IOP Publishing

Key Nuclear Reaction Experiments
Discoveries and consequences
Hans Paetz gen. Schieck

Chapter 5

Halo nuclei and farewell to simple radius systematics

At the 'rims' of the valley of stability (the neutron or proton *driplines*) there are a number of nuclei that have much larger radii than expected from the systematics. ^{11}B has about the same radius as ^{208}Pb.

The first experimental evidence of halos in 1985 was deviations of reaction cross sections from the systematics expected as described in chapter 4 in light nuclear isotopes far from the valley of stability, such as ^{11}Li [1, 2], see also [3, 4]. With increased interaction (absorption) radii between nuclei in heavy-ion reactions, narrower momentum distributions of breakup fragments in such reactions have also been observed [5].

Also the deuteron has an extreme rms radius of about 3.4 fm. In all cases the nuclei seem to have a halo of weakly bound neutrons (or protons), which surrounds a more strongly bound core. Different cores are possible, i.e. in addition to the strongly bound α making ^{5}He and ^{6}He one- and two-neutron halo nuclei, ^{9}Be also forms some type of core. Generally indications of halo structures are—among others—exceptionally large cross sections in heavy-ion reactions, narrower momentum distributions of the nucleons in the nuclei and larger radii, as compared to the $A^{1/3}$ law. Figure 5.1 shows the low-mass portion of the chart of nuclides where halo nuclei have been found. The scientific interest in halo nuclei is manifold. They were among the first where the driplines have been reached. The results show that the shell structures established for the valley of stability can be extended to 'exotic' nuclei, but with modifications of the closed shells, i.e. with new magic numbers emerging. The low mass numbers invite application of microscopic theories such as Faddeev(–Yakubowsky), no-core–shell models, Green's function Monte Carlo and other approaches to test nuclear forces, e.g. three-body forces, or effective-field theory approaches. Impressive results have been obtained by such *ab initio* calculations, see e.g. [6, 7].

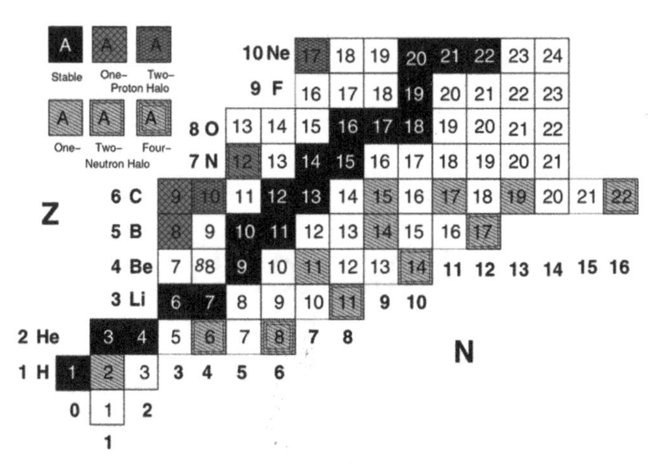

Figure 5.1. Halo nuclei at the driplines of the chart of nuclides. The fields with numbers only belong to unstable nuclides. For properties such as lifetimes, see e.g. the chart of nuclides of [8].

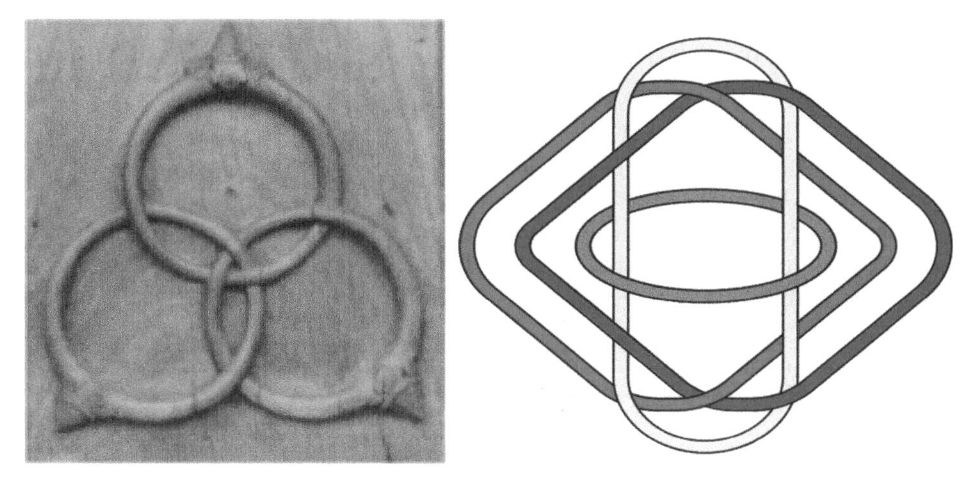

Figure 5.2. Coat of arms and symbol of the Renaissance Borromean family (and other north Italian families like the Sforzas) at their castle on the Borromean island Isola Bella in the Lago Maggiore, Italy (left). Schematic super-Borromean or Brunnian link structure of four intertwined rings as a model of a nucleus such as ^{10}C. Reproduced from [9], see also [10] (right).

A special role is played by the so-called *Borromean* nuclei, i.e. those that consist of a core plus two weakly bound neutrons at large radii, and for which any of the two-particle subsystems are unbound (example: ^{4}He + n + n). They can be treated by well-established three-body methods. Their name is derived from the three intertwined Borromean rings that fall apart when one ring is removed and hold together only when united, see figure 5.2. An especially exotic case is that of the nucleus ^{10}C which seems to have a four-body cluster structure $\alpha + \alpha + \mathrm{p} + \mathrm{p}$ and has been called a *super-Borromean* or *Brunnian link* nucleus because all

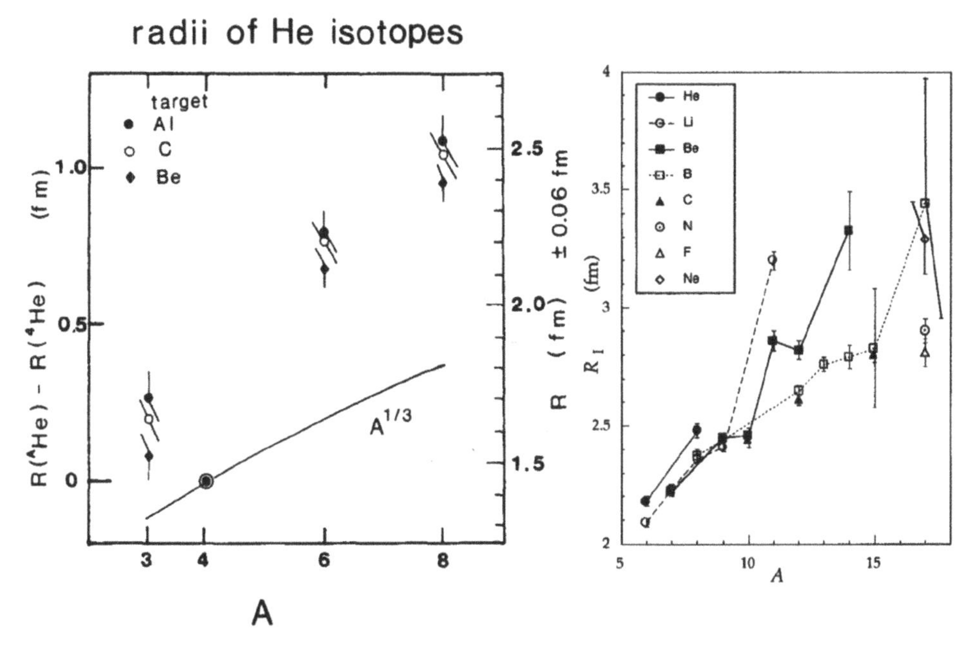

Figure 5.3. (Left) The rms radii of He isotopes determined from interaction cross sections compared to the $A^{1/3}$ systematics. Reproduced with permission from [2]. Copyright 1996 IOP Publishing. (Right) For a number of other nuclei see [11]. Reproduced with permission from [11]. Copyright 1996 Springer.

subsystems (e.g. ^{5}He, ^{8}Be, or 2p) are unbound whenever one component is removed, see e.g. [10]. Many nucleosynthesis processes pass through nuclei that are neutron rich or neutron poor and are not well known. Thus, for nuclear astrophysics, a better understanding of all these reactions and their reaction rates is essential.

Since we deal with unstable (radioactive) nuclei the 'radioactive ion beam (RIB)' facilities, which are in operation or being developed, are especially suited for their investigation. These facilities collect, focus and accelerate nuclear reaction products in order to use them as projectiles in reactions. Two main operating principles have been developed: the 'isotope separator online' (ISOL) principle and in-flight fragmentation' (IFF) with the REX-ISOLDE facility at CERN as an example. The separated and accelerated radioactive beams are often used on lighter targets such as hydrogen in small-angle *inverse* kinematics.

The first and key experiments that led to the discovery of halo nuclei were performed by Tanihata *et al* [1, 2, 4]. For the comparison of interaction cross sections of different isotopes, e.g. as function of isospin, the simple ansatz $\sigma_I = \pi[R_I + r_0 A^{1/3}]^2$ was assumed where A is the mass number of the stable target nucleus and R_I the radius of the unstable projectile to be determined via the measured interaction cross section σ_I. The experiments were performed at the Berkeley BEVALAC accelerator. A selection of these early results is shown in figures 5.3 and 5.4. Also, requiring some model calculations, the matter density

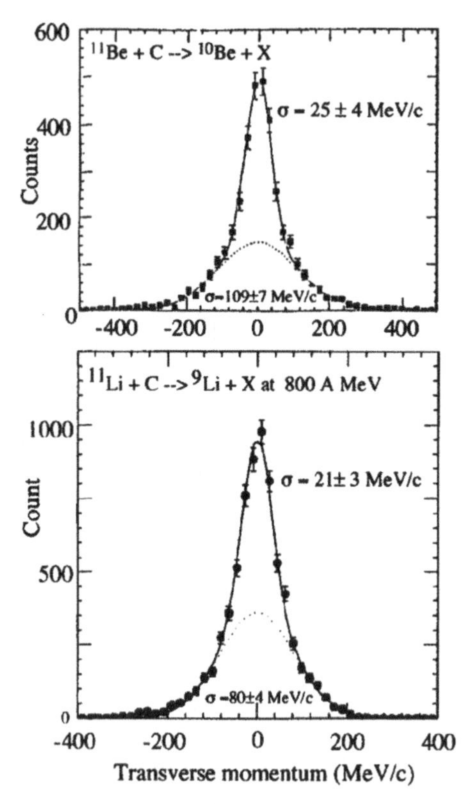

Figure 5.4. Transverse momentum distributions in ^{11}Be and ^{11}Li, i.e. of the cores ^{10}Be and ^{9}Li, against one or two halo neutron(s) that are smaller than the core nuclei. Reproduced with permission from [11]. Copyright 1996 IOP Publishing.

distributions of neutron halo nuclei were deduced in [1] and found to be much wider than the proton distributions as well as those of the cores or, e.g., stable nuclei. A later measurement [5] confirmed this and is shown in figure 5.5 and with a modern RIB set-up in figure 5.6. Figures 5.3, 5.4 and 5.5 show the characteristic properties of halo nuclei:

- Their interaction cross sections are larger than for non-halo nuclei.
- They have (matter) radii, determined by halo neutrons, which are larger than predicted from the usual $A^{1/3}$ systematics valid for stable nuclei, for the cores of halo nuclides, as well as for charge radii that are very little affected by the neutron halos, see e.g. [13] and figure 4.13.
- Their density distributions reach further out than usual.
- In agreement with this they show narrower momentum distributions of the breakup fragments of the halo nuclei (one example: ^{19}C → ^{18}C + n, compared to ^{17}C → ^{16}C + n).

Already in the first halo experiments a number of halo nuclei have been identified, among them: ^{11}Li, ^{11}Be, ^{14}Be, ^{17}B and ^{17}Ne, all except ^{11}Be being two-neutron

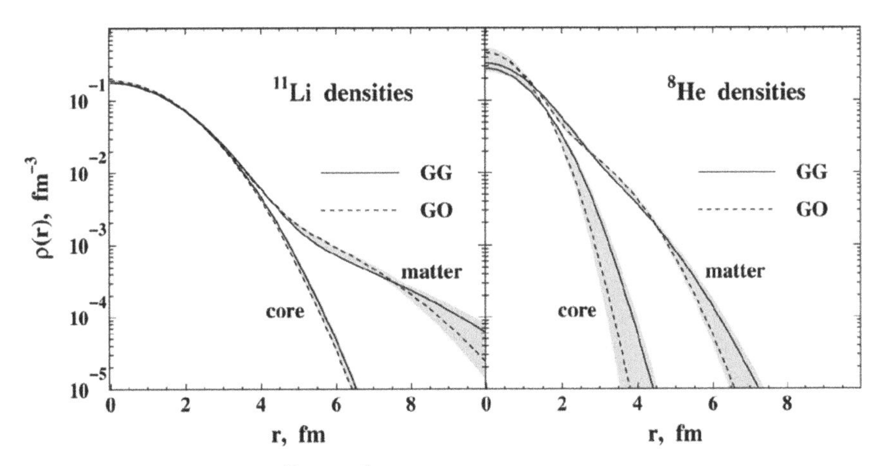

Figure 5.5. Density distributions in ^{11}Li and ^{8}He, typical two or four neutron halo nuclei. Reproduced with permission from [5]. Copyright 2006 Elsevier.

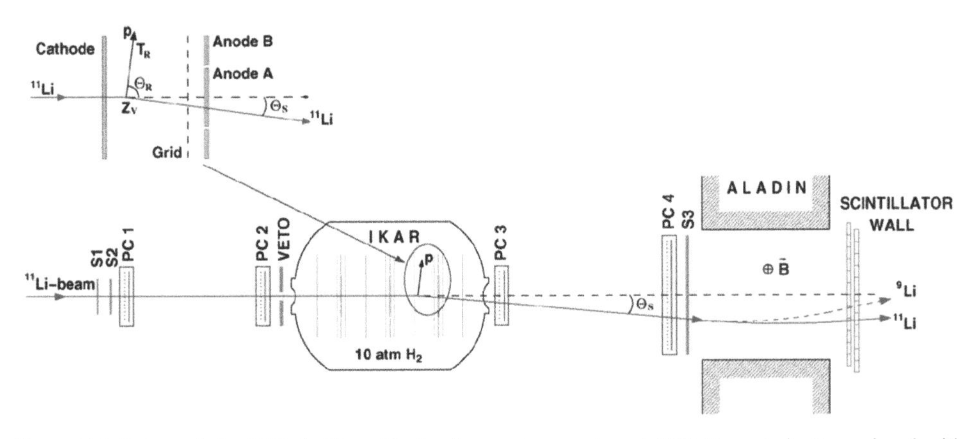

Figure 5.6. Set-up of the RIB facility with the fragment separator at GSI Darmstadt. Reproduced with permission from [5]. Copyright 2006 Elsevier.

halo nuclei. Systematic studies of nuclear (matter) radii using radioactive beams, mainly by elastic scattering on protons, were performed and in the framework of the Glauber model yielded systematic rms matter radii [4], see also [11, 12].

^{11}Li has become the 'cornerstone' of halo research and thus of recent *ab initio* calculations of lighter nuclei. By comparison with different models a three-body structure of a ^{9}Li core plus a two-neutron halo with the extremely low two-neutron separation energy of ≈300 keV has been identified, see [14, 15].

A recent discovery of a halo nucleus is that of ^{22}C [16] which showed an increased reaction cross section and an rms radius of $r_{\mathrm{rms}} = 5.4 \pm 0.9$ fm, both larger than expected from the usual systematics. For a recent survey of the status of halo nuclei, see e.g. [17].

Bibliography

[1] Tanihata I, Hamagaki H, Hashimoto O, Shida Y, Yoshikawa N, Sugimoto K, Yamakawa O, Kobayashi T and Takahashi N 1985 *Phys. Rev. Lett.* **55** 2676
Tanihata I *et al* 1985 *Phys. Lett.* B **160** 380
[2] Tanihata I 1985 *Hyperfine Interact.* **21** 251
[3] Krieger A *et al* 2012 *Phys. Rev. Lett.* **108** 142501
[4] Ozawa A *et al* 2001 *Nucl. Phys.* A **691** 599
Ozawa A, Suzuki T and Tanihata I 2001 *Nucl. Phys.* A **693** 32
[5] Dobrovolsky A V *et al* 2006 *Nucl. Phys.* A **766** 1
[6] Dean D J 2007 *Phys. Today* **60** 11, 48
[7] Pieper S C and Wiringa R B 2001 *Ann. Rev. Nucl. Part. Sci.* **51** 53
[8] National Nuclear Data Center NNDC, Brookhaven National Laboratory 2012 http://www.nndc.bnl.gov/chart
[9] Wikipedia 2015 *Brunnian Link* public domain https://en.wikipedia.org/wiki/Brunnian_link
[10] Curtis N *et al* 2008 *J. Phys.: Conf. Ser.* **111** 012022
[11] Tanihata I 1996 *J. Phys. G: Nucl. Part. Phys.* **22** 157
[12] Tanihata I 1991 *Nucl. Phys.* A **522** 275c
[13] Nörtershäuser W *et al* 2009 *Phys. Rev. Lett.* **102** 062503
[14] Hansen P G, Jensen A S and Jonson B 1995 *Ann. Rev. Nucl. Part. Sci.* **45** 591
[15] Hansen P G and Jonson B 1985 *Europhys. Lett.* **4** 409
[16] Tanaka K *et al* 2010 *Phys. Rev. Lett.* **104** 062701
[17] Simon H 2013 *Phys. Scr.* **T152** 014024

IOP Publishing

Key Nuclear Reaction Experiments

Discoveries and consequences

Hans Paetz gen. Schieck

Chapter 6

The particle zoo

Although the basic structure of nuclei became clear after the discovery of the proton and the neutron around 1935, the neutrino as a product of β decay was still a postulate as well as Hideki Yukawa's exchange particle, later called the pion. Then, new (elementary?) particles (such as the muon $\mu^{\pm}$ and the positron e^{+}) found in cosmic rays started to open the entirely new and complex world of particle physics. The parallel development of accelerators immediately suggested the artificial production of these and the search for more such particles at higher and higher energies commensurate with increasing masses of these particles, a trend that continues up to this day (examples are the $W^{\pm}$ and Z^{0} exchange bosons, the top quark t and the Higgs boson). A very large number of particles have been identified and classified according to their properties as leptons, hadrons, hyperons, baryons, fermions or bosons, see the most recent tabulations of the Particle Data Group [35]. A more fundamental order was given to this zoo by the discovery of the building blocks of six leptons, six quarks and their antiparticles, in the framework of the standard model (SM).

6.1 The pion

Yukawa in 1935 postulated an exchange particle with a mass of $\approx$130 MeV c^{-2}, commensurate with the range of the (hadronic) nuclear force of about 1.4 fm. For a time the muon, found in 1932 in cosmic radiation, was mistaken as this particle, but—being a heavy lepton—it did not have the expected properties. Only in 1947 was the pion detected as a component of cosmic rays [27–29] in photoplates, see figure 6.1, decaying as

$$\pi^{\pm} \rightarrow \mu^{\pm} + \nu_{\mu}. \tag{6.1}$$

The sign of its charge could not be determined due to the lack of a strong magnetic field. The pion was identified with a mass of $\approx 300\, m_{\mathrm{e}}$ by its shorter lifetime (range) and the ionization grain density being higher than that of the muon. All muon tracks

doi:10.1088/978-0-7503-1173-1ch6

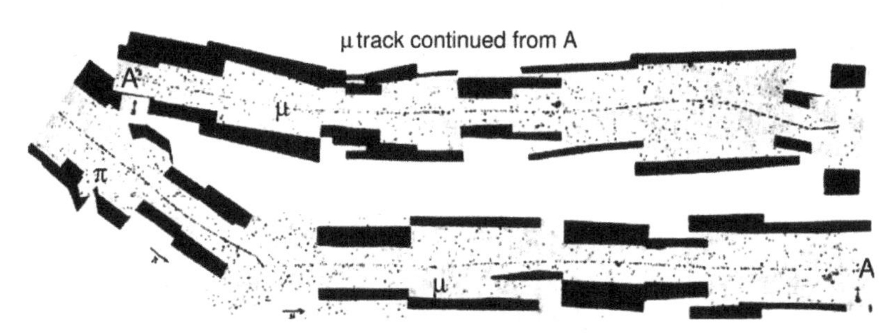

Figure 6.1. Photoplate tracks of the decay of a cosmic pion into a muon and an invisible muonic (anti) neutrino, as seen through a microscope. The typical length of the muon track is ≈0.61 mm. Adapted with permission from [28]. Copyright 1947 Nature Publishing Group.

observed were of about equal length, corresponding to a pion decay at rest. Sometimes the pion decay would be accompanied by a number of strong hadronic tracks of short range, i.e. α particles, etc, forming a 'star', which can be taken as evidence of a hadronic interaction, i.e. through the strong interaction between the pion and an emulsion nucleus, in contrast to the pion decay that is a weak-interaction process. Figure 6.2 shows such an event. Soon after this discovery—in 1948—the artificial production of pions in accelerators of sufficient energy, e.g. the 184″ synchrocyclotron at Berkeley, was successful [13, 22, 26]. A carbon target positioned inside the magnetic field of the synchrocyclotron was bombarded with protons of energies starting near the production threshold just below 200 MeV. Positive and negative pions were deflected by the magnetic field in opposite directions and registered by two stacks of photoplates. These showed not only microscopic tracks of pions but also their decay into muons (with typical track lengths of 0.6 mm) and the β decay of the muons. The properties of pions could now be studied as a function of their energies, e.g by scattering them from hydrogen. The interaction strength and the shape of the differential cross sections pointed to the strong interaction as the one keeping nuclei together (against the repulsive Coulomb force between the protons). The spin and parity of the pions were determined to be $J = 0^-$ (they are *pseudoscalars*), their isospin $T = 1$, meaning that they come as an isobaric triplet $\pi^\pm$, π^0.

The study of the scattering of pions from protons proved to be particularly fruitful. Resonances in excitation functions were indications of the formation of intermediate states, corresponding to excited states of the nucleons. The most prominent is the excitation of the Δ resonance. Because of the high energies involved and thus the high number of decay channels these excited states are rather short-lived, i.e. the resonances are quite broad. The Δ has $J^\pi = 3/2^+$ and isospin 3/2 meaning that it comes as an iso-quartet. These quantum numbers require a spin flip whereas another such resonance, the *Roper* resonance, has the same quantum numbers as the proton and can be interpreted as an excited proton without change of internal structure. In this way a large number of 'particles' could be created, such as the *baryons*, which encompass the nucleons and the *hyperons* Δ, Λ, Σ, Ξ and Ω, the latter four carrying the new quantum number of *strangeness*. In the

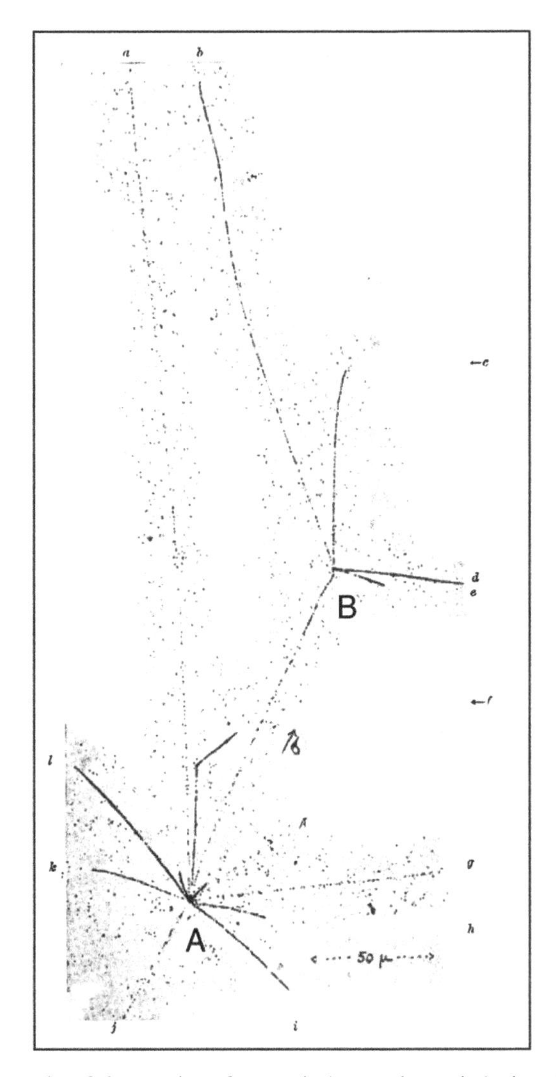

Figure 6.2. Photoplate tracks of the reaction of a cosmic (assumed negative) pion after slowing down and being captured in an atomic shell and cascading down into the K shell before being destroyed in a hadronic interaction with a nucleus forming a 'star'. Stars (e.g. at vertex A) may consist of α particles from nuclear interactions (strong tracks), protons and pions, which can decay into muons with long, weakly ionized tracks, and also undergo another nuclear interaction (vertex B). Reproduced with permission from [29]. Copyright 1947 Nature Publishing Group.

constituent-quark model all baryons are made of three quarks u and d, quark–antiquark pairs, allowing for s- and $\bar{s}$-quarks and gluons. In similar experiments a number of mesons were discovered: The scalar mesons: ρ, ω, ϕ and K^* and the pseudoscalar mesons π, η, η' and K. All mesons consist of quark–antiquark pairs. The multitude of particles is often called 'the particle zoo', into which only the assignment of quark combinations brought perfect order.

6.2 The first production of the antiproton in a nuclear reaction

Paul Dirac in 1928 created a relativistic form of the wave equation, the Dirac equation with energy eigenvalues for a free electron

$$E = \pm\left(p^2c^2 + m_e c^2\right)^{1/2}$$

[19]. He associated the seemingly unphysical negative solution with a hole in an otherwise filled sea of negative-energy states and thus with the electron's antiparticle, the positron. After the discovery of the positron in cosmic rays in a cloud chamber by Carl Anderson in 1932 [3, 4] and the observation of positive and negative muons by J C Street and E C Stevenson [37] and Anderson and Seth Neddermeyer in 1937/1938 [5, 6], the question naturally arose of the existence of antiparticles for every particle which could only be answered through experiments. Simultaneously the concepts of lepton-number and baryon-number conservation were developed.

An antiparticle has the same properties as the associated particle, except for charge-related properties (charge, magnetic moment). Antiparticles annihilate when interacting with *ordinary* matter, i.e. their energy is radiated as neutral radiation such as photons or π^0. Antiprotons had not been identified in cosmic rays before and the Bevalac synchrotron accelerator at Berkeley was specifically designed to find the antiproton.

The conservation laws (energy and momentum conservation, and charge conjugation) require that when bombarding a nuclear target with (high-energy) protons and with an antiproton in the exit channel the antiparticle must be one of a particle–antiparticle pair (i.e. $\mathrm{p} + \bar{\mathrm{p}}$). In addition ejectile and recoil particles will appear, thus four particles are emitted, i.e. the reaction had to be

$$\mathrm{p} + \mathrm{p} \rightarrow \mathrm{p} + \mathrm{p} + \mathrm{p} + \bar{\mathrm{p}},$$

in order to obey baryon number conservation for $B = 2$. Applying relativistic kinematics for a fixed target particle with mass m_2 we need a laboratory energy of incident beam particles with mass m_1 for the energy available in the center-of-mass (c.m.) system of $E_{\mathrm{c.m.}}$

$$T_{\mathrm{lab}} = \frac{E_{\mathrm{c.m.}}^2}{2m_2c^2} + E_{\mathrm{c.m.}}\left(1 + \frac{m_1}{m_2}\right).$$

The threshold for the creation of the pair $\mathrm{p} + \bar{\mathrm{p}}$ is defined by $E_{\mathrm{c.m.}} = 2m_\mathrm{p}c^2$. The proton laboratory energy threshold for antiproton production on a fixed target of free protons is thus

$$T_{\mathrm{thr}} = 6 \cdot m_\mathrm{p}c^2 = 5.638\ \mathrm{GeV}. \tag{6.2}$$

Such an energy could only be reached with synchrotron accelerators. In this case, the Bevatron at Berkeley, designed for a maximum beam momentum of 6.3 GeV c^{-1}, corresponding to an energy of 5.4 GeV, was used to create the first antiprotons in 1955 [14].

In the actual experiment a copper target was used. Because of the binding of the protons in the heavier target the Fermi momentum of the nucleons could be exploited to add some relative energy to the projectile–target system with the effect of lowering the threshold energy for $\bar{\mathrm{p}}$ production. With the approximate factor

$$1 - p_{\mathrm{F}}/M_{\mathrm{p}}c \tag{6.3}$$

the threshold momentum is lowered to a value of about $4.8\,\mathrm{GeV\,c^{-1}}$. Figure 6.3 shows schematically the set-up to produce antiprotons. How was the unique

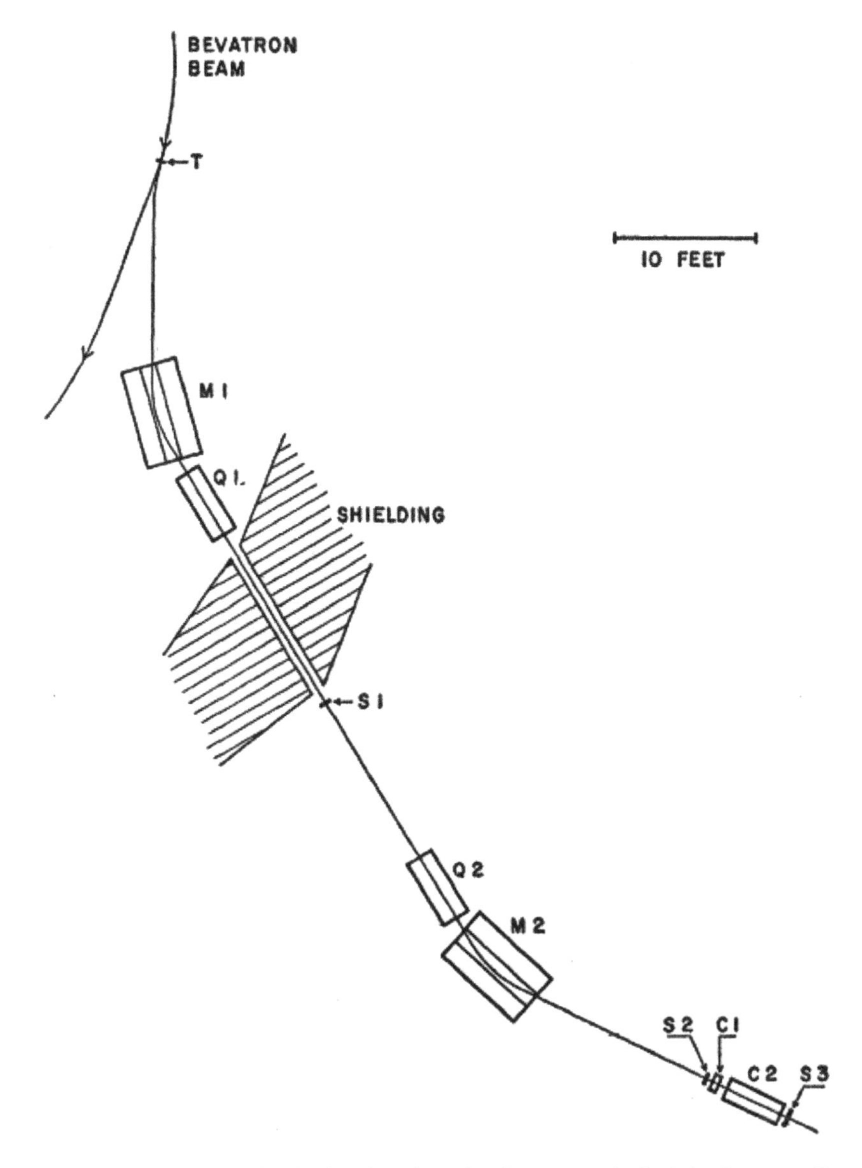

Figure 6.3. Experimental set-up for the first detection of antiprotons at the Berkeley Bevatron. Reproduced with permission from [14]. Copyright 1955 American Physical Society.

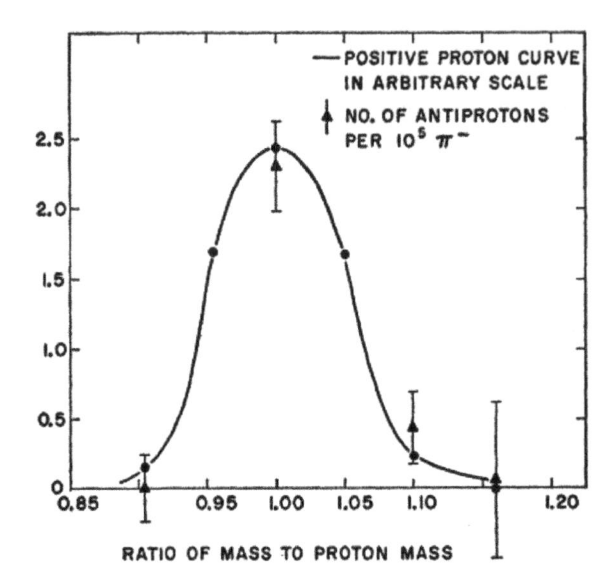

Figure 6.4. Transmission of the detector arrangement for negative particles with a mass of $\approx m_p$, i.e. of antiprotons $\bar{p}$. Reproduced with permission from [14]. Copyright 1955 American Physical Society.

identification of the antiprotons achieved? In view of an enormous background of (negative) pions the following measurements were taken:

- Use was made of the negative charge leading to a deflection and extraction opposite to the deflection of the protons in the magnetic field of the synchrotron.
- Near the threshold the antiprotons move with half the beam velocity, i.e. with momentum 1.19 GeV c^{-1} to which the magnet system is tuned.
- The time of flight between two scintillation detectors about 10 m apart is measured by delayed coincidence techniques to obtain the velocity and thus obtain an idea of the mass.
- A combination of three Čerenkov detectors (which are velocity-sensitive detectors), one a $\beta = v/c$ threshold detector the other with a narrow β window, were used to register the antiprotons only and reject pions and vice versa, leading to a strong background suppression.

The system was tuned to transmit particles of mass m_p, and a small variation of the beam momentum yielded a transmission curve for negative particles centered about the proton mass, as shown in figure 6.4. Figure 6.5 shows the production rate of antiprotons, relative to the rate of pions as function of the proton laboratory energy starting from the expected threshold at about 4.2 GeV.

6.3 Discovery of the (electron) neutrino

Wolfgang Pauli formulated the *neutrino hypothesis* in 1930 [30] as an answer to open questions concerning β decay. The continuous electron spectrum and the spins of the particles called for a third decay particle. Enrico Fermi formulated his theory of the weak interaction in 1934 [21]. Nevertheless the first neutrino (in its electron

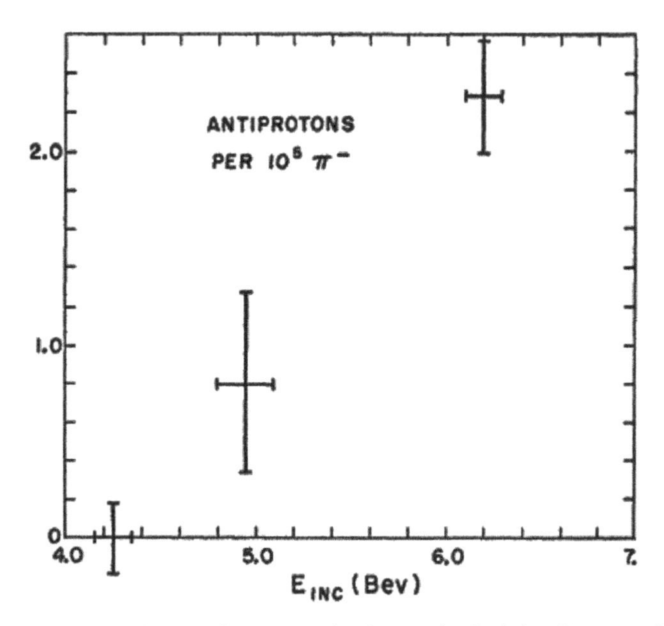

Figure 6.5. Excitation curve of the antiproton production at the Berkeley Bevatron. Reproduced with permission from [14]. Copyright 1955 American Physical Society.

flavor) was only detected directly in experiments in 1953 by Frederick Reines and Clyde L Cowan [16, 31–33]. Only then was β decay satisfactorily explained and conservation laws such as lepton-number conservation could be postulated.

The experiment was based on the reaction

$$\bar{\nu} + \mathrm{p} \rightarrow \mathrm{n} + \beta^{+} \tag{6.4}$$

(an inverse β decay). The two-particle exit channel is a condition for sharp energies (line spectra) facilitating their detection. Figure 6.6 depicts the scheme of the detection set-up. Antineutrinos arise in large numbers from power reactors. Due to the neutron excess of the primary fission product nuclei these undergo β decay with the emission of antineutrinos—according to baryon- and lepton-number conservation. Fluxes on the order of $10^{13}\,\mathrm{cm}^{-2}\,\mathrm{s}^{-1}$ are available, which—even with the very small weak-interaction cross section of typically $10^{-43}\,\mathrm{cm}^2$—provides well-measurable event rates of $\approx 5 \cdot 10^{-3}\,\mathrm{s}^{-1}$. The different parts of the experiment are:

- The neutrons from the reaction (6.4) are moderated (time scale: several hundred μs) and captured in a liquid target consisting of $CdCl_2$ in H_2O thereby emitting γ rays that are registered in two large liquid scintillator counters (in anti-coincidence for background reduction). ^{113}Cd (with an abundance of 12.26% in natural Cd) has a very large absorption cross section for thermal neutrons ($\sigma = 63\,600$ b at $E_\mathrm{n} = 0.18$ eV).
- The positrons lose energy, are finally stopped in the target and annihilate into two 511 keV annihilation γ quanta emitted in opposite directions in two scintillator tanks where they are registered in coincidence. The coincidence

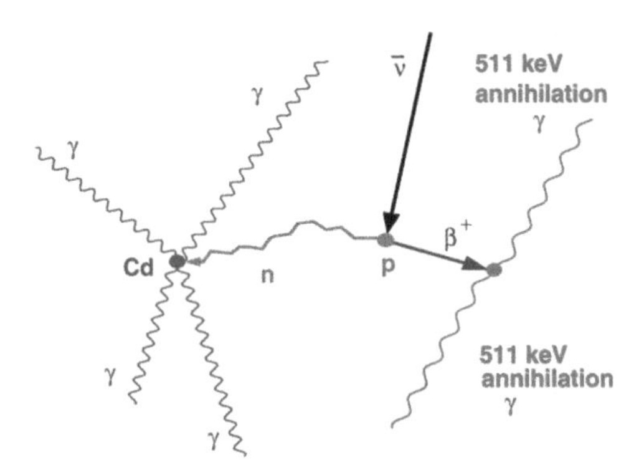

Figure 6.6. Antineutrino reaction and detection scheme of the neutron-capture γ from Cd and positron annihilation 511 keV γ in a suitable liquid scintillator containing a neutron moderator (water) and a Cd compound (e.g. $CdCl_2$).

and anti-coincidence requirements lead to an elaborate construction for the large target and scintillator tanks.

- The two (related) coincidence events are measured in an additional time-delayed coincidence with variable delay time around the moderation time of the neutrons in the target. With 'true' events an average cross section of several measurements was

$$\sigma = 11 \pm 2.6 \cdot 10^{44}\,\mathrm{cm}^2. \tag{6.5}$$

The authors announced their success in finding the neutrino in June 1956 in a famous telegram to Pauli at ETH Zürich [34].

A later experiment that became famous for the measurement of solar neutrinos was based on the reaction

$$\nu_e + {}^{37}\mathrm{Cl} \rightarrow {}^{37}\mathrm{Ar} + \mathrm{e}^- \tag{6.6}$$

[15, 17, 18]. In this experiment the small cross section was compensated by a very large detector containing many tons of chlorine compounds such as CCl_4, see figure 6.7. The experiment ran continuously for about 25 years. It turned out that the ^{37}Ar produced could be extracted from the target liquid quantitatively and identified by its electron capture (EC) β decay

$$^{37}\mathrm{Ar} + \mathrm{e}^- \rightarrow {}^{37}\mathrm{Cl} + \nu, \tag{6.7}$$

which is accompanied by the emission of three to five atomic K-shell Auger electrons. Figure 6.8 shows the miniature proportional counter used for measuring the activity of the ^{37}Ar β decay. The signals were selected by an imposed pulse rise-time condition, thus reducing the background. Great care was observed in the selection of all materials to reduce any background rate substantially below that of true solar neutrino events ($\approx$1 event per week). The set-up was tested at the Brookhaven reactor and a power-plant reactor and then ran for many years 1.5 km underground in the Homestake mine in South Dakota to avoid background events from terrestrial sources.

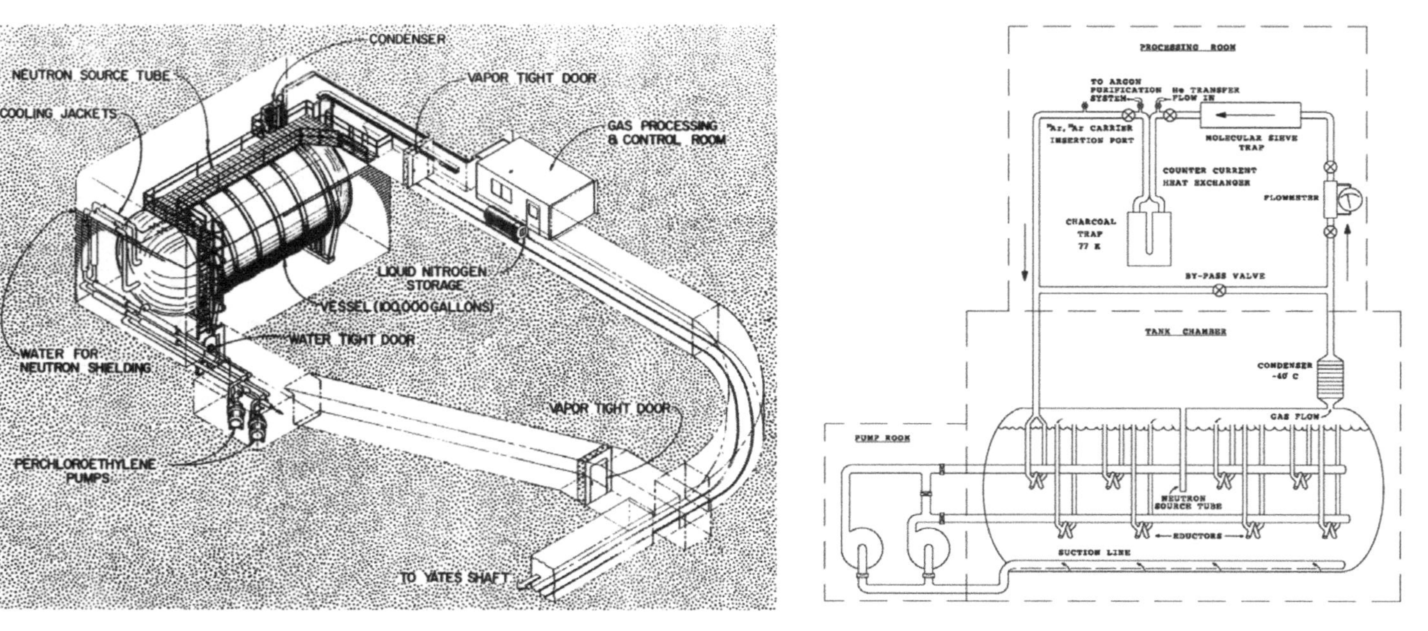

Figure 6.7. A view and schematic of the tank for the unique detection of solar neutrinos. Reproduced with permission from [15]. Copyright 1998 IOP Publishing.

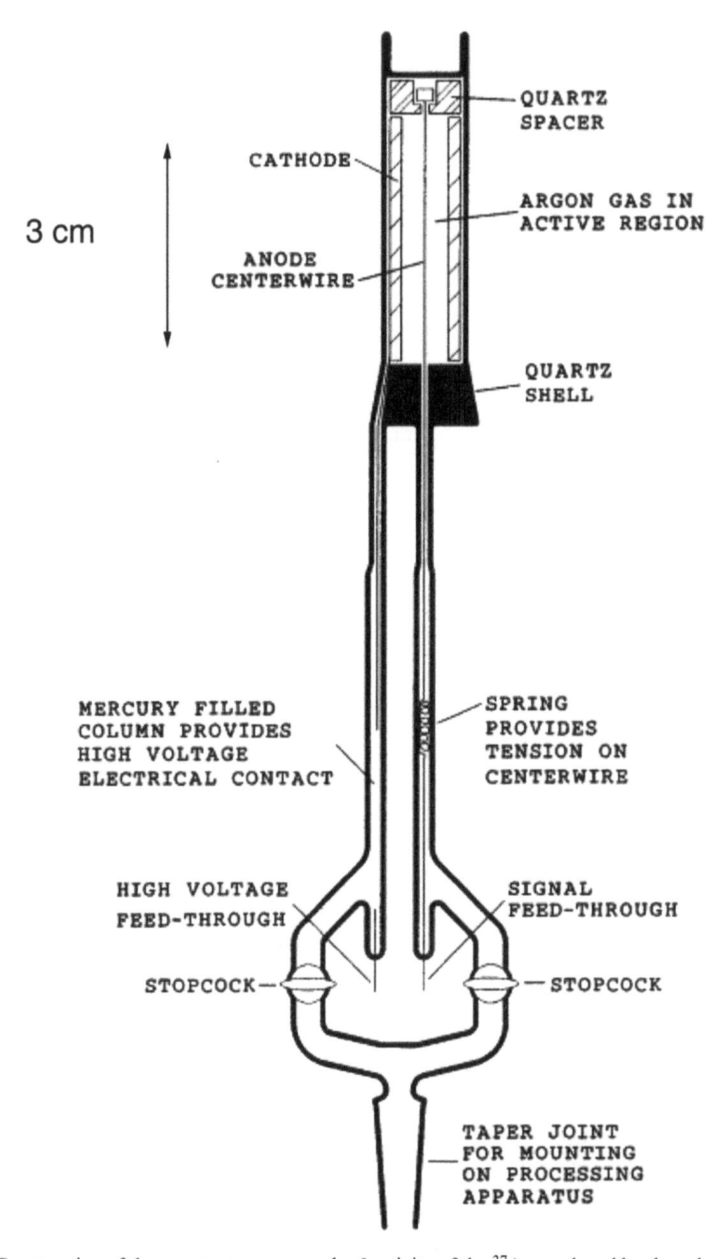

Figure 6.8. Construction of the counter to measure the β activity of the ^{37}Ar produced by the solar neutrinos and extracted from the chlorine compound. Reproduced with permission from [15]. Copyright 1998 IOP Publishing.

The results of these measurements were manifold:

- The experiment proved that antineutrinos are emitted from the fusion reactions in the Sun's interior.
- The neutrino and antineutrino are different particles: the cross section measured of $\sigma < 0.9 \cdot 10^{-45}$ cm^2 per Cl atom is incompatible with that expected if the neutrino and antineutrino were indistinguishable ('Majorana neutrino').

A consequence would be that neutrinoless double-β decay would be allowed and that neutrinos must have mass. However, in contrast to the double-β decay with emission of two antineutrinos, e.g. in the decay

$$^{106}_{48}\text{Cd} \rightarrow {}^{106}_{46}\text{Pd} + 2\text{e}^{+} + 2\bar{\nu}_{\text{e}}, \tag{6.8}$$

this process has not been found. Neutrino oscillations have been detected, however:

- At the Super-Kamiokande detector in Japan evidence of oscillations of cosmic-ray neutrinos was found in 1998.
- In 2002 oscillations of solar electron antineutrinos to another flavor were seen by the Sudbury Neutrino Observatory in Ontario 2 km underground, thus solving completely the 'solar-neutrino puzzle'.
- Oscillations of antineutrinos from 22 different nuclear power plants at different distances were detected by the KamLAND experiment in Japan in 2002 [8, 20]. The detector was a massive kiloton liquid scintillator viewed by nearly 2000 photomultiplier tubes in the Kamioka mine.

The intensity of the solar neutrino flux as measured by the Davis experiment was only about one third of that expected from model calculations of solar models using all known parameters of the Sun as inputs [7, 15, 36]. The solution of this 'solar neutrino puzzle' was that electron neutrinos have mass and thus can 'oscillate', i.e. periodically transform into one or both of the other neutrino flavors ν_μ and ν_τ (the evidence, for cosmological as well as particle physics reasons, is very strong that there exist only three 'families' of leptons—likewise for the hadron families of quarks). With the existence of these oscillations the finite mass of the neutrinos is a fact—it remains, however, open as to what the values of the masses of each of the three neutrino flavors are.

6.4 Quasi-elastic electron scattering—excited nucleons and the particle zoo

Electron scattering off nuclei (e.g. protons) with low momentum transfer can resolve these as a whole, but not smaller constituents such as quarks.

The spectra of (weakly) inelastic electron scattering from the proton at relatively high energies show, in addition to the elastic peak, a number of excited states (as in nuclear scattering spectra), corresponding to excited nucleon states (nucleon resonances). This follows from the cross-section behavior with q^2, the transferred momentum squared, which is similar for ground- and excited-state transitions, see e.g. [11] and figure 4.7 of chapter 4.

6.5 Deep-inelastic lepton scattering—partons inside hadrons

Once higher electron energies became available, the finer structures of nucleons and nuclei could be explored. The key experiment was performed by M Breidenbach, J I Friedman, H W Kendall and R E Taylor (Nobel Prize 1990), see [12, 38] around 1966. Figure 6.9 shows the double-differential deep-inelastic and the differential elastic cross sections as functions of q^2 (corresponding to a combination of incident energy and scattering angle).

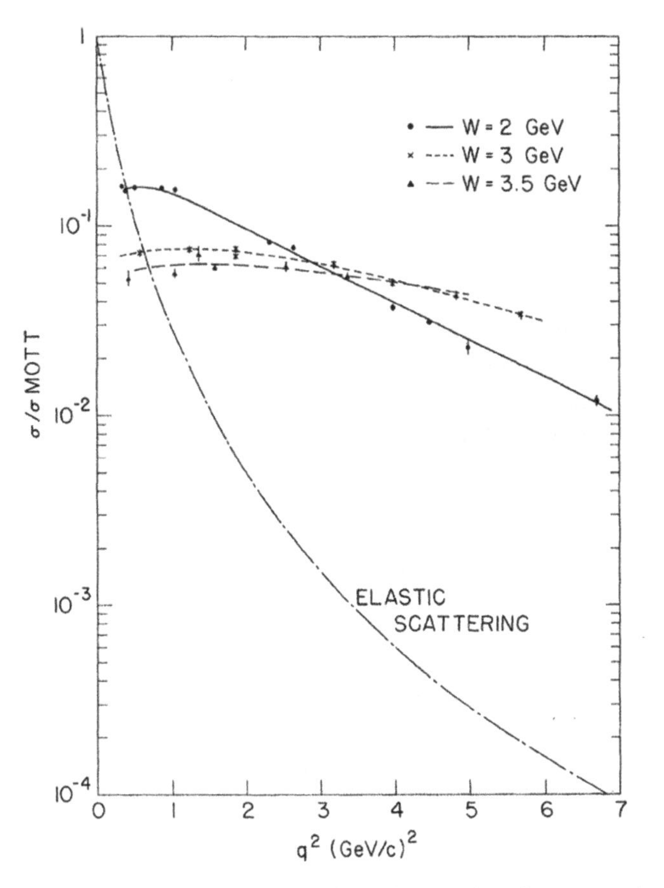

Figure 6.9. Double-differential cross section of deep-inelastic scattering of 500 MeV electrons and elastic scattering versus q^2, the square of the transferred momentum. The most remarkable feature is the very different q^2 dependence of the elastic and the inelastic cross sections. The data are plotted relative to the (point) Mott cross sections. The lines are to guide the eye only. Reproduced with permission from [12]. Copyright 1969 American Physical Society.

The experiment yielded clear evidence of 'parton' structures inside the nucleons that had all the properties of the quarks, postulated theoretically in 1964 by Murray Gell-Mann, who coined the term 'quark' after James Joyce's *Finnegan's Wake*, and George Zweig [23, 41]. They recognized that all known hadrons of the time could be fitted nicely into a scheme (the 'eightfold way' [24]) of combinations of only three different particles with certain properties. Most conspicuous was the necessity of their having fractional charges (±1/3e or ±2/3e), but also an additional degree of freedom called 'color'. The existence of strange particles required the introduction of a strange quark (the s-quark) for which the discovery of the J/Ψ meson as an s–anti-s combination brought final evidence, as well as two more types of quarks, the bottom (b) and finally the top quark (t) (and their antiparticles) which closed the scheme. The evidence for the vector bosons ('gauge bosons') mediating the strong interaction between quarks via virtual exchange—and thus responsible for the stability of nuclei—was first discovered around 1976 in electron–positron collisions at the PETRA accelerator at DESY/Hamburg and the particles were called 'gluons'.

The characteristics of these results were the occurrence of 'three-jet' events, i.e. the emission of quark–antiquark pairs plus a sideways single jet from a neutral particle. The quarks as well as the gluons are 'confined', i.e. cannot be liberated to appear as free particles, but show up as jets of hadrons instead. This is also expressed by the idea of the quarks and gluons carrying the property of color and the postulate that free particles have to be color-neutral, i.e. do not carry color. Further information on the properties and history of all particles can be found in the regularly updated reviews of particle properties, see [35].

The 'scaling' behavior found in deep-inelastic electron scattering (as in the classical Rutherford scattering) proves that these constituents are apparently point-like and therefore truly 'elementary' (to our current knowledge!).

The scattering of neutrinos from nuclei required a number of special preparations. In order to form an 'intense' beam of neutrinos relativistic kinematics had to be

Figure 6.10. The Big European Bubble Chamber (3.7 m) at CERN. A strong magnetic field of 3.5 T produced by a superconducting magnet enables the determination of charge and momentum of the charged particles. Copyright 1974 CERN.

applied, which causes reaction products to be emitted into narrow forward cones. The neutrinos were produced in the decay of pions and muons that accompany the reactions of high-energy protons with suitable solid targets. The invention of the *Van der Meer neutrino horn* [25, 39] appreciably increased the neutrino beam density. In this magnetic focusing device the beam of pions whose decay produces a beam of neutrinos is strongly focused in the forward direction with the same effect on the neutrinos. Only through this focusing did the more recent long-distance neutrino-oscillation experiments become possible.

The search for neutrino oscillations (transformation into other neutrino flavors as a function of the traveling distances or flight times) has led to very large detectors filled with neutrino-sensitive liquids and scintillators and surrounded by numerous photomultipliers (Super-Kamiokande, Sudbury Observatory, KamLAND, Gran Sasso, etc) looking for solar, cosmic and reactor-produced neutrinos. Thus, neutrino oscillations have been discovered and the solar-neutrino puzzle has been solved.

Deep-inelastic scattering of neutrinos from protons at CERN around 1979 (see e.g. [1] and figures 6.10 and 6.11) as well as of muons [2, 9] and electrons [40] gave some evidence that inside the proton (and of course, other hadrons) not only quarks

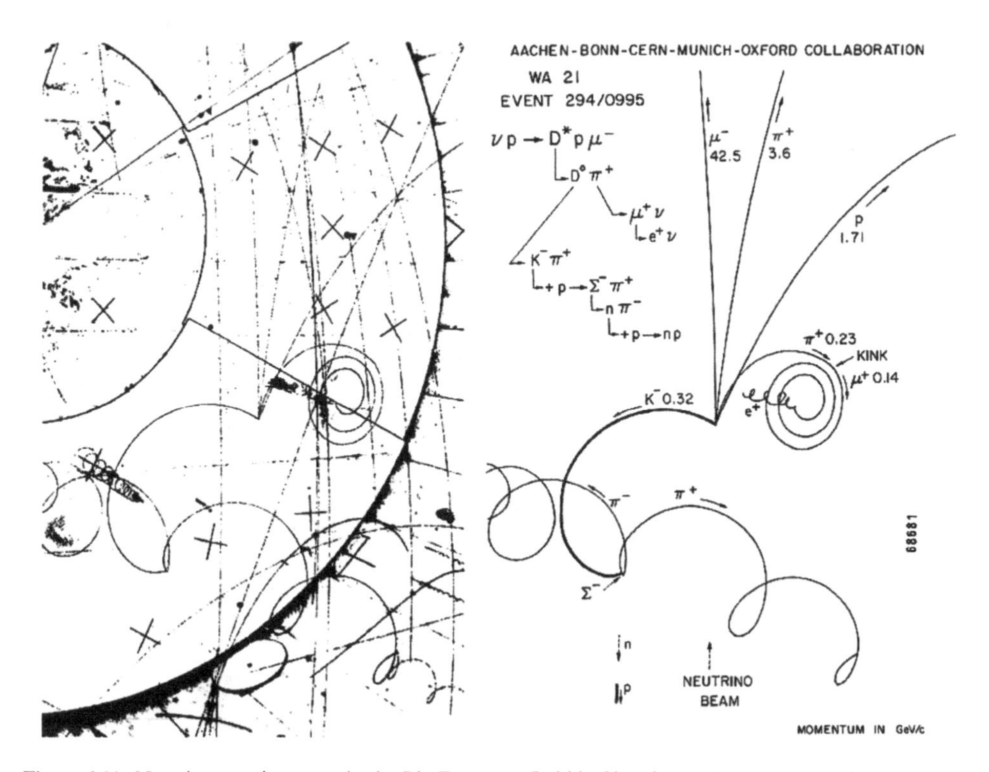

Figure 6.11. Neutrino reaction event in the Big European Bubble Chamber at CERN. A neutrino, produced by a 350 GeV proton beam at the Super Proton Synchrotron accelerator, entering from the bottom, reacts with a proton of the H_2 filling the chamber and produces the indicated sequence of events, among them a charmed D meson transforming into a singly produced strange K^-. Reproduced with permission from [10]. Copyright 1979 Elsevier.

but some other electrically neutral 'material' exists which was dubbed 'gluons'. Only half of the momentum of the proton was carried by the quarks, the other half was obviously carried by the gluons as shown by the 'structure functions' (similar to the low-energy form factors) of the nucleons. More detailed experiments in which polarized particles were used as projectile and/or targets yielded 'spin-structure functions' exhibiting the different contributions to the nucleon's spin $1/2\hbar$. This global spin quantum number could not be explained by the addition of the spins of the constituent quarks alone ('the spin crisis') but also required contributions from the gluon spin and from orbital angular momenta.

Bibliography

[1] Aguilar-Benitez M *et al* (Particle Data Group) 1992 *Phys. Rev.* D **45** 1
[2] Amaudruz P *et al* 1992 *Phys. Lett.* B **295** 159
[3] Anderson C D 1932 *Science* **76** 238
[4] Anderson C D 1933 *Phys. Rev.* **43** 491
[5] Anderson C D and Neddermeyer S 1937 *Phys. Rev.* **51** 884
[6] Anderson C D and Neddermeyer S 1938 *Phys. Rev.* **54** 88
[7] Bahcall J N and Pinsonneault M H 1995 *Rev. Mod. Phys.* **67** 781
[8] Bahcall J N, Gonzalez-Garcia M and Peña-Garay C 2003 *J. High Energy Phys.* JHEP02(2003)009
[9] Benvenuti A C *et al* 1990 *Phys. Lett.* B **237** 592
[10] Blietschau J *et al* 1979 *Phys. Lett.* B **86** 108
[11] Bloom D *et al* 1969 *Phys. Rev. Lett.* **23** 931
[12] Breidenbach M, Friedman J I, Kendall H W, Bloom E D, Coward D H, DeStaebler H, Drees J, Mo L W and Taylor R E 1969 *Phys. Rev. Lett.* **23** 935
[13] Burfening J, Gardner E and Lattes C G M 1948 *Phys. Rev.* **75** 382
[14] Chamberlain O, Segrè E, Wiegand C and Ypsilantis T 1955 *Phys. Rev.* **100** 947
[15] Cleveland B T, Daily T, Davis Jr R, Distel J R, Lande K, Lee C K, Wildenhain P and Ullman J 1998 *Astrophys. J.* **496** 505
[16] Cowan C L and Reines F 1956 *Science* **124** 103
[17] Davis R 1955 *Phys. Rev.* **97** 766
[18] Davis R 1964 *Phys. Rev. Lett.* **12** 303
[19] Dirac P A M 1928 *Proc. R. Soc.* A **117** 610
[20] Eguchi K *et al* (Kamland Collaboration) 2003 *Phys. Rev. Lett.* **90** 021802
[21] Fermi E 1934 *Z. Physik* **88** 161
[22] Gardner E and Lattes C M G 1948 *Science* **107** 270
[23] Gell-Mann M 1964 *Phys. Lett.* **8** 214
[24] Gell-Mann M and Ne'eman Y 1966 *The Eightfold Way* (New York: Benjamin)
[25] Giesch M, Kuiper B, Langeseth B, Van der Meer S, Neet D, Plass G, Pluym G and de Raad B 1963 *Nucl. Instrum. Methods* **20** 58
[26] Jones S B and White R S 1949 *Phys. Rev.* **78** 12
[27] Lattes C M G, Muirhead H, Occhialini G P S and Powell C F 1947 *Nature* **159** 694
[28] Lattes C M G, Occhialini G P S and Powell C F 1947 *Nature* **160** 453
[29] Lattes C M G, Occhialini G P S and Powell C F 1947 *Nature* **160** 486
[30] Pauli W 1930 Letter to the 'Radioaktiven Damen und Herren' ('radioactive ladies and gentlemen') *Tübingen meeting* (*4 December 1930*)

[31] Reines F and Cowan C L 1953 *Phys. Rev.* **92** 830

[32] Reines F and Cowan C L 1959 *Phys. Rev.* **113** 273

[33] Reines F 1960 *Ann. Rev. Nucl. Sci.* **10** 1

[34] Reines F 1995 *Nobel Lecture*, The Nobel Foundation (Stockholm) and (Singapore: World Scientific)

[35] Particle Data Group 2008 Review of particle properties *Rev. Mod. Phys.* **80** 633

[36] Sackman J, Boothroyd A I and Fowler W A 1990 *Astrophys. J.* **360** 727

[37] Street J C and Stevenson E C 1937 *Phys. Rev.* **52** 1003

[38] Taylor R E 1967 *Proc. Int. Symp. on Electron and Photon Interactions at High Energies (1967, Stanford, CA)*

[39] Van der Meer S 1961 *CERN Report* 61–7

[40] Whitlow L *et al* 1992 *Phys. Lett.* B **282** 475

[41] Zweig G *CERN Report* No. 8182/TH401 (unpublished)

Chapter 7

Discovery of the neutron (nuclear kinematics, etc)

As early as 1920 Ernest Rutherford had hypothesized a neutral particle of about the same mass as a proton, capable of explaining the shortcomings of the then current ideas on nuclear structure. However, no experimental evidence supported this idea. Walther Bothe and Herbert Becker [1] bombarded a number of elements with α particles and observed a penetrating radiation of very high energies that they interpreted as γ rays. With Be and B in particular, unusually high intensities were observed, see figure 7.1. Irène Curie and Frédéric Joliot investigated the absorption of this radiation in lead and observed a half-value thickness of 4.7 cm for Pb which would correspond to an initial γ energy of 15–20 MeV for the Be rays, energies never observed before in natural γ emissions. The subsequent insertion of either non-hydrogenous and hydrogenous materials after the Pb filter had the effect of a strong and unexpected increase in count rate for the second category. They interpreted the ejected protons as being due to a *nuclear Compton effect* from these γ [6]. The correctly measured energy of the protons from the Be source of 5.7 MeV could only be reconciled with an initial photon energy of 55 MeV. A more detailed description of these dramatic steps towards the discovery of the neutron can be found in [9].

7.1 Chadwick's discovery

Chadwick discovered the neutron in 1932 [2] by correctly identifying the energetic radiation emitted from the reaction $\alpha + {}^{9}_{4}\mathrm{Be} \rightarrow {}^{12}_{6}\mathrm{C} + {}^{1}_{0}\mathrm{n}$ (induced by α from a Po source). The recoil energies transferred to the protons and ^{14}N nuclei of the filling gas of the ionization chamber were measured to be 5.7 MeV and 1.6 MeV, respectively.

The value of the neutron mass (in u) obtained by Chadwick was 1.0067. From range measurements of protons ejected as recoils from hydrogenous material only a crude estimate of the neutron energies and its mass could be deduced. However,

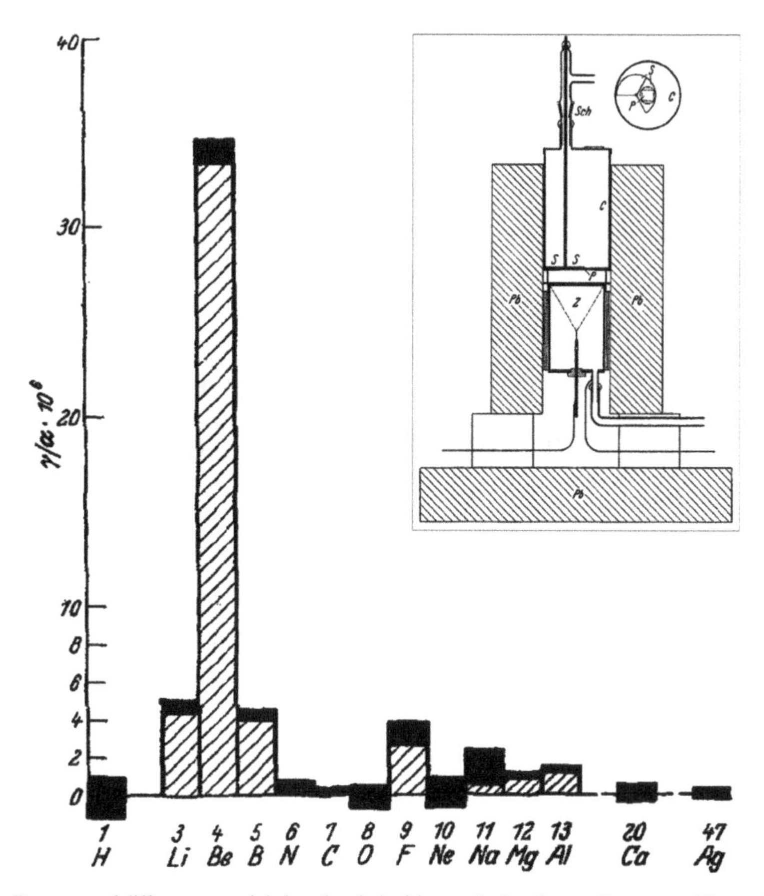

Figure 7.1. Response of different materials bombarded with α radiation from a Po source. The emitted strong radiation was believed to be γ but was later identified to be neutrons. The inset shows the apparatus used (Z: Geiger counter, P: Po α source, C: evacuated vessel, S: target substance, Pb: lead shielding against possible background radiation). Modified with permission from [1]. Copyright 1930 Springer.

using the same apparatus but observing the reaction $\alpha + {}^{11}_{5}\text{B} \rightarrow {}^{14}_{7}\text{N} + {}^{1}_{0}\text{n}$ because of better-known mass values, he obtained the mass of the neutron within the limits between 1.005 and 1.008. Today's best value is 1.008664904 ± 0.000000014 u.

Other nuclear physicists (among them the Curies) had erroneously interpreted the energetic radiation as γ radiation with energies up to 50 MeV, transferring recoil energy by some Compton-like scattering process to the protons or ${}^{14}\text{N}$ nuclei. However, such energies of γ transitions cannot occur in nuclei. Only the assumption of a neutral particle with a mass near that of the proton appeared consistent with all observations. The reaction in question was the production of recoil protons from the hydrogenous material in paraffin by neutrons produced in Be by the ${}^{9}\text{Be}(\alpha,\,\text{n}){}^{12}\text{C}$ reaction with a Q-value of 5.7 MeV. The additional insertion of thin aluminum sheets of variable thickness after the paraffin led to a decrease of the pulse height of the ionization chamber signal commensurate with the energy loss of protons in Al, but not electrons. Figure 7.2 shows the extremely simple experimental set-up used by Chadwick. A kinematical calculation, i.e. the application of the classical conservation

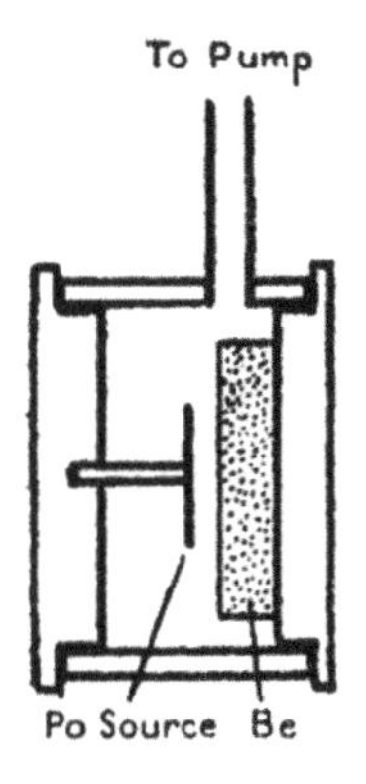

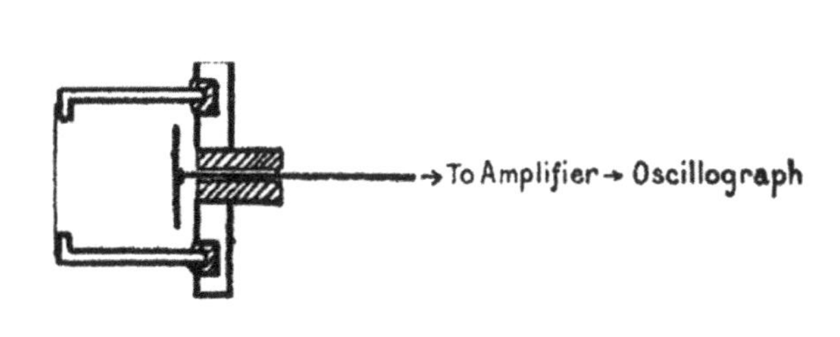

Figure 7.2. Apparatus used by Chadwick to discover the neutron. On the left is the Po–Be α source in a vacuum, producing an energetic radiation later identified as neutrons. On the right is the ionization chamber as a detector for protons. With a plate of paraffin between the two the count rate increased substantially. From [2].

laws of energy and momentum and comparison of the neutron–proton collisions in paraffin ($E_p = 5.7$ MeV) with those of a nitrogen-containing material ($E_N = 1.5$ MeV) led Chadwick to the approximate mass of the neutron

$$\frac{E_p}{E_N} = \frac{m_p}{m_n}\left(\frac{m_n + m_N}{m_n + m_p}\right)^2 = \frac{5.7}{1.5}$$

with the rough result of

$$m_n \approx 1.15 m_p$$

which—in view of the experimental uncertainties of about 10%—is quite good. Chadwick received the Nobel Prize in 1935.

7.2 The structure of nuclei and the role of neutrons

Initially the measured mass of the neutron led to the assumption of the neutron being a (quasi-)bound proton–electron system, but very shortly this idea was dismissed in favor of the neutron being an (elementary) particle of its own.

Immediate consequences of the discovery were:

- For the first time the model of nuclei as being composed of protons and neutrons, the existence and 'construction' of isotopes, the periodic table and the chart of nuclides became unambiguous. It was proposed independently by Werner Heisenberg [7] and Dmitri Ivanenko [8].
- The neutron, due to its electric neutrality, proved to be an ideal projectile to penetrate nuclei and to perform all kinds of nuclear reactions at all energies, including the induced fission of heavy nuclei and the creation of heavier isotopes (in the laboratory and in nucleosynthesis). Neutron multiplication allowing a chain reaction after fission is the basis of nuclear reactors for energy production as well as for the atomic bomb.
- Neutron physics has become a very important discipline with far-reaching results. Thus, many different methods had to be developed for the production of neutrons over an enormous range of energies from ultracold ($<\mu$eV) to

many GeV using nuclear reactions in radioactive sources and accelerators, spallation sources and nuclear reactors. On the other hand, the detection techniques became refined, again using nuclear reactions, leading to charged particles, fission chambers, activation, time-of-flight methods, etc.

- The similarity of the properties of protons and neutrons led Heisenberg [7] to postulate the symmetry of *charge independence* and the conservation quantity for which E P Wigner later coined the term isospin [11], see also section 12.3. Both have been very important in nuclear and particle physics, also because their symmetry later proved to be slightly broken.
- The antineutron was first produced and studied at Berkeley around 1956 (see [3, 4]).
- The β decay

$$\mathrm{n} \rightarrow \mathrm{p} + \mathrm{e}^- + \bar{\nu}_\mathrm{e} \tag{7.1}$$

is a prototype process caused by the weak interaction. The idea of the existence of the particle class of leptons, especially of neutrinos (antineutrinos), is intimately connected with this process. This decay was already measured in 1914 by Chadwick displaying a continuous β spectrum and thus could not be a two-particle decay, provided the kinematics of energy and momentum conservation holds. Only between 1930, when Wolfgang Pauli tentatively formulated the neutrino hypothesis [10], 1934, when Enrico Fermi created his theory of β decay and weak interaction, and 1956, when the neutrino was detected directly by Frederick Reines and Clyde L Cowan [5], were β decays satisfactorily explained.

- The neutron allows a large number of applications ranging from neutron-activation analysis, the creation of medically required isotopes by neutron-capture reactions in reactors, neutron radiography complementing x-ray studies in materials analysis, to the study of biological structures.

Bibliography

[1] Bothe W and Becker H 1930 *Z. Phys.* **66** 289

[2] Chadwick J 1932 *Nature* **129** 3252
Chadwick J 1932 *Proc. R. Soc.* A **136** 692

[3] Chamberlain O, Segrè E, Wiegand E and Ypsilantis T *Nature* **177** 11

[4] Cork B, Lambertson G R, Piccioni O and Wenzel W A 1956 *Phys. Rev.* **104** 1193

[5] Cowan C L and Reines F 1956 *Science* **124** 103

[6] Curie I and Joliot F 1932 *C. R. Acad. Sci., Paris* **93** 273

[7] Heisenberg W 1932 *Z. Phys.* **77** 1
Heisenberg W 1932 *Z. Phys.* **78** 156

[8] Iwanenko D 1932 *Nature* **129** 798

[9] Marmier P and Sheldon E 1970 *Physics of Nuclei and Particles* vol 2 (New York: Academic)

[10] Pauli W 1930 Letter to the 'Radioaktiven Damen und Herren' ('radioactive ladies and gentlemen') *Tübingen Meeting* (*4 December 1930*)

[11] Wigner E 1937 *Phys. Rev.* **51** 106

Chapter 8

The first precise determination of the neutron mass and the binding energy of the deuteron

After the discovery of the neutron by James Chadwick and the implications for the structure of nuclei it was essential to measure its properties more precisely, among them its spin and mass. The deuteron had just been discovered by Urey *et al* [8] in 1932 by optical spectroscopy and its approximate mass was determined from the recoil factor $(1 + m_e/m_N)^{-1}$ in the energies of spectral lines of optical spectra (m_e, m_N are the masses of the electron and the isotope investigated, respectively). Two principal methods could be applied to measure the binding energy of the deuteron.

8.1 The photonuclear disintegration of the deuteron

The photo-disintegration of the deuteron

$$\gamma + \mathrm{d} \rightarrow \mathrm{p} + \mathrm{n}$$

(analogous to the atomic *photo effect*) was used by Chadwick and Goldhaber [3] to measure m_d. A strong γ source of Th C″ (today ^{208}Tl) with $E_\gamma = 2.615$ MeV was used to incite the reaction. D_2 gas was used as the filling gas in an ionization chamber and the energy of protons stopped in the chamber was measured from the height of the oscilloscope traces produced. A test with a Ra C (today ^{214}Bi) source with a γ emission line with $E_\gamma = 1.764$ MeV did not produce any protons. Thus, the threshold for photo-disintegration had to be between these two values. The pulse heights of the proton traces allowed the deduction of a kinetic energy of $E_{kin} \approx 250$ keV.

doi:10.1088/978-0-7503-1173-1ch8

The neutrons must have had the same energy, hence the *binding energy* (BE) of the deuteron

$$\mathrm{BE(d)} \approx 2.1\ \mathrm{MeV}.$$

With Albert Einstein's mass–energy relation $m_d c^2 = m_p c^2 + m_n c^2 - \mathrm{BE(d)}$ we obtain the mass of the neutron

$$m_n = m_d - m_p + \mathrm{BE(d)}/c^2.$$

Bainbridge [2] measured the masses of the proton and deuteron and with these values and error bars the mass of the neutron was

$$m_n = 1.0080 \pm 0.0005\ \mathrm{u}.$$

With later improved methods m_n was determined more precisely by measuring the disintegration threshold using x-ray bremsstrahlung from an electron accelerator and registration of the neutrons in a BF_3 gas-filled proportional counter that makes use of the reaction $^{10}B + n \rightarrow {}^{7}Li + \alpha + 2.78\ \mathrm{MeV}$. One of these 'precision' experiments was performed in 1959 by R C Mobley and R A Laubenstein [6] at the pressurized ANL accelerator. ANL had the specialty that it could accelerate ions to ground potential and simultaneously accelerate electrons to a target inside the tank on high voltage where the experiment with bremsstrahlung γ could be performed. The ion beam served to determine the energy of the electrons indirectly via a calibrated nuclear reaction.

This resulted in

$$\mathrm{BE(d)} = 2.227(3)\ \mathrm{MeV} \qquad \text{and} \qquad m_n = 1.008982(3)\ \mathrm{u}$$

The experimental arrangement with the electron accelerator as the x-ray source is shown in figure 8.1, the target and detector set-up in figure 8.2. The γ energy (i.e. the maximum energy of the bremsstrahlung spectrum) could be varied by varying the electron energy to pass the threshold for the reaction

$$^{2}\mathrm{H}(\gamma, \mathrm{n})^{1}\mathrm{H}.$$

It turned out that the intensity increase above the threshold was quadratic such that a linear dependence of the square root of the yield would allow a precise extrapolation to the threshold energy corresponding to the binding energy of the deuteron. The method and results are shown in figure 8.3.

8.2 Neutron–proton capture

The inverse reaction (*neutron–proton capture*)

$$\mathrm{n} + \mathrm{p} \rightarrow \mathrm{d} + \gamma$$

is another possibility to measure BE(d) by measuring the energy of the γ produced with thermal neutrons. Before 1977 a number of such measurements had been performed,

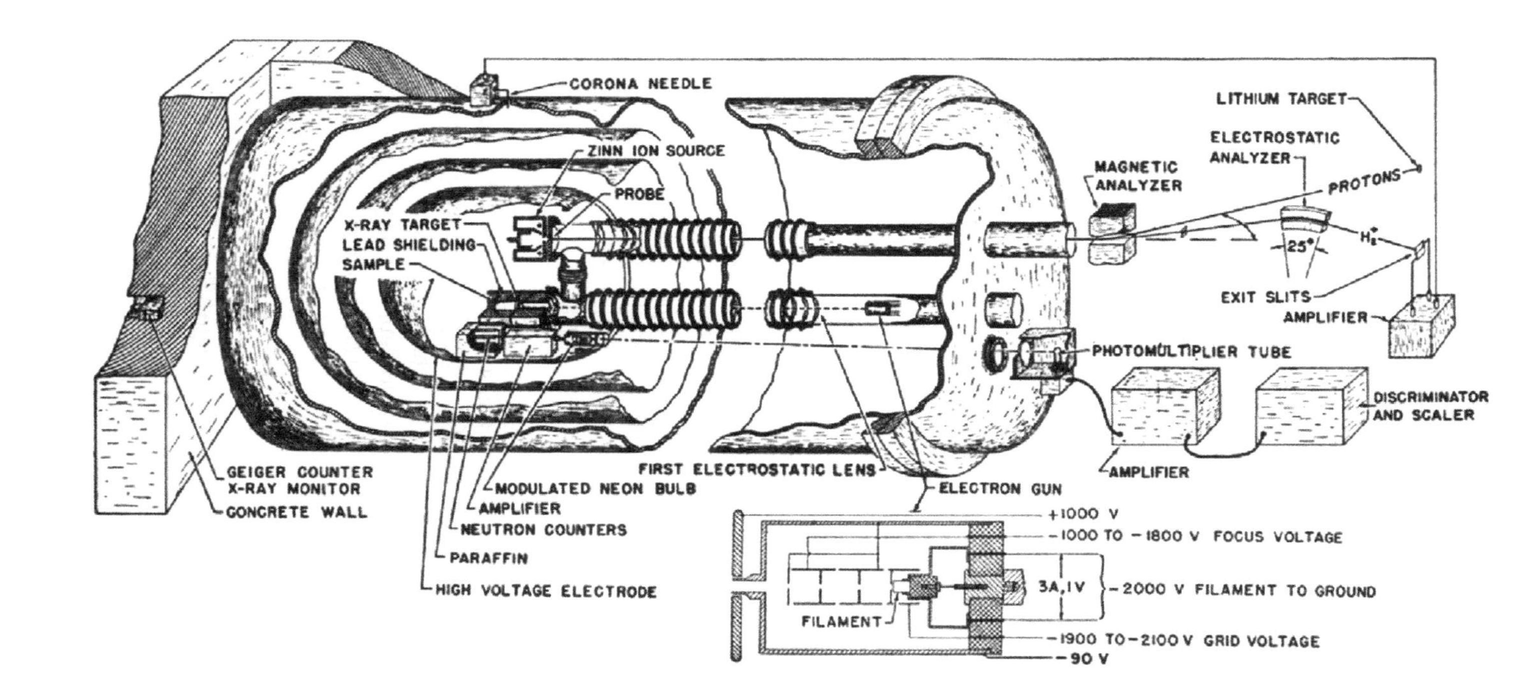

Figure 8.1. A view of the bremsstrahlung source, an electron accelerator used in the experiment of [6]. Reproduced with permission from [6]. Copyright 1950 American Physical Society.

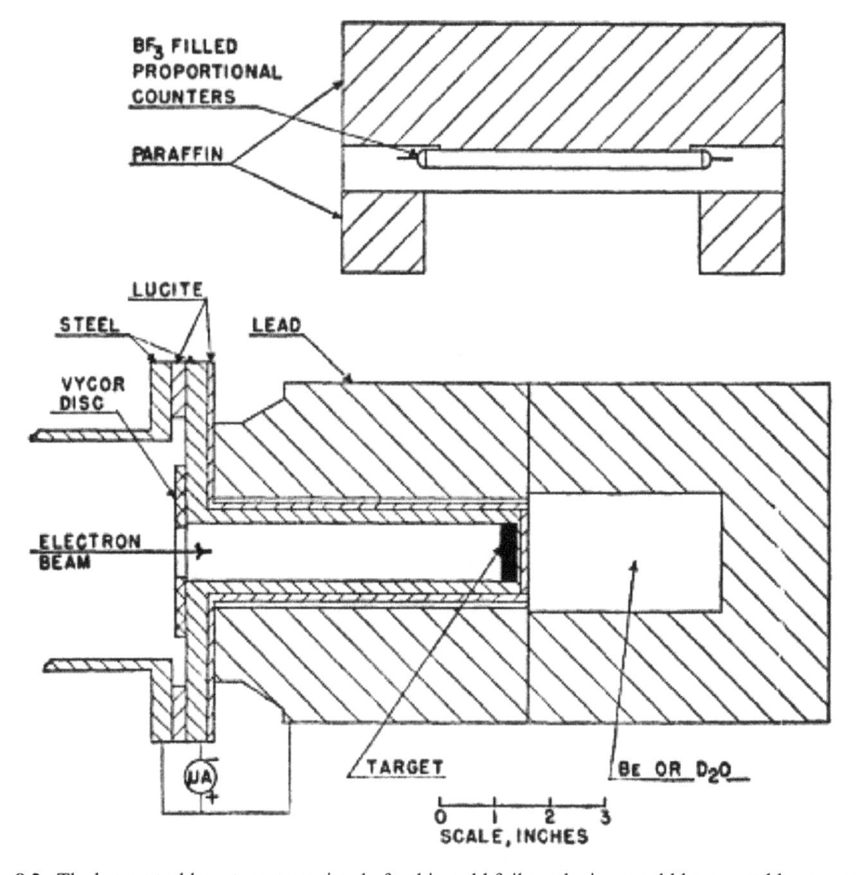

Figure 8.2. The bremsstrahlung target consisted of a thin gold foil producing a gold bremsstrahlung spectrum, the reaction target was deuterium in the form of heavy water in a thin-walled container and the neutron detector was a $^{10}BF_3$ proportional counter [6]. Reproduced with permission from [6]. Copyright 1950 American Physical Society.

but gave conflicting results. Using new calibration standards for γ energies [5] Van der Leun *et al* [9] published the results of a high-precision experiment using an n-type intrinsic Ge detector and another Ge(Li) detector for the γ spectroscopy. The set-up of the experiment is shown in figure 8.4 and figure 8.5 displays the relevant spectrum of the intrinsic Ge detector. The neutrons were produced by a ^{241}Am $+^{9}$Be source and thermalized in a paraffin cylinder. The result of

$$\mathrm{BE(d)} = 2224575 \pm 9\ \mathrm{eV}$$

in comparison to earlier data is shown in figure 8.6. The accepted best values at present ([7] and [1]) are

$$\mathrm{BE(d)} = 2.224566\ \mathrm{MeV} \qquad \text{and} \qquad m_{\mathrm{n}} = 1.00866491600\ \mathrm{u}.$$

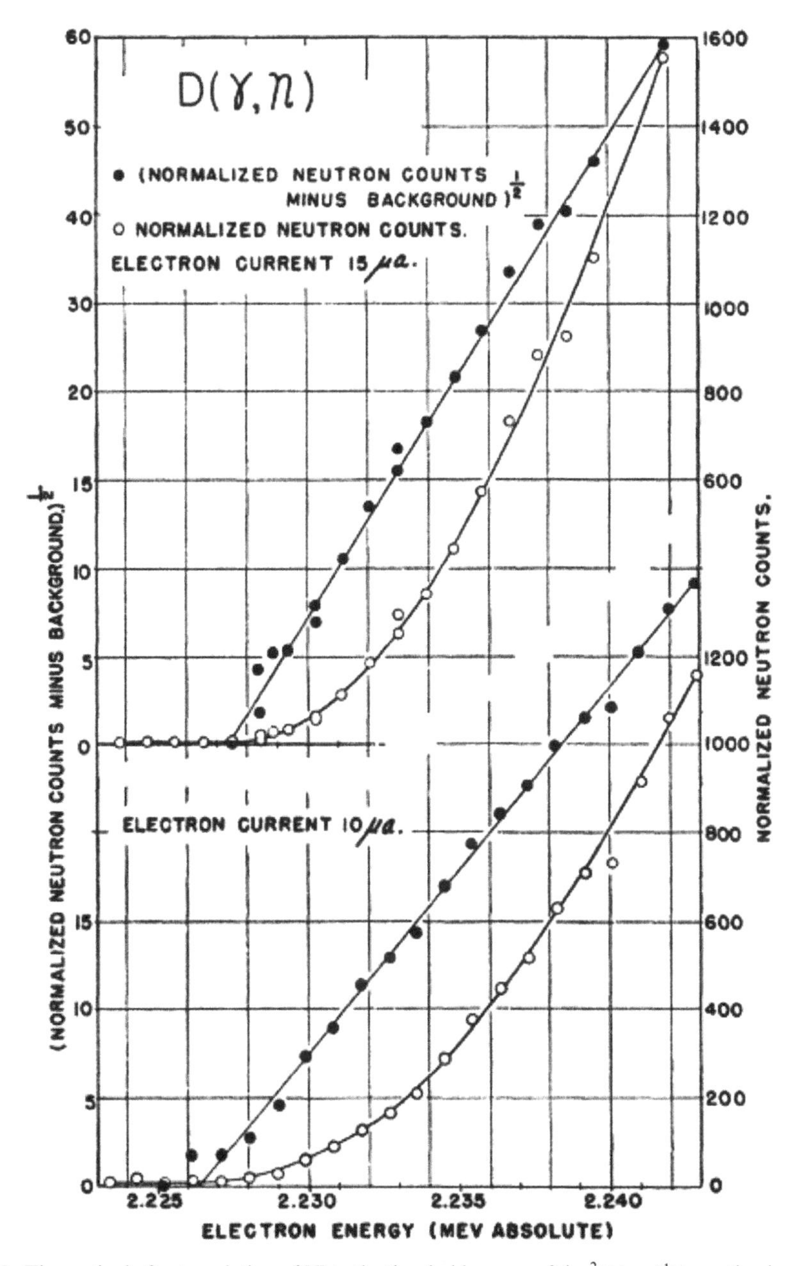

Figure 8.3. The method of extrapolation of [6] to the threshold energy of the $^{2}H(\gamma, n)^{1}H$ reaction is explained. Reproduced with permission from [6]. Copyright 1950 American Physical Society.

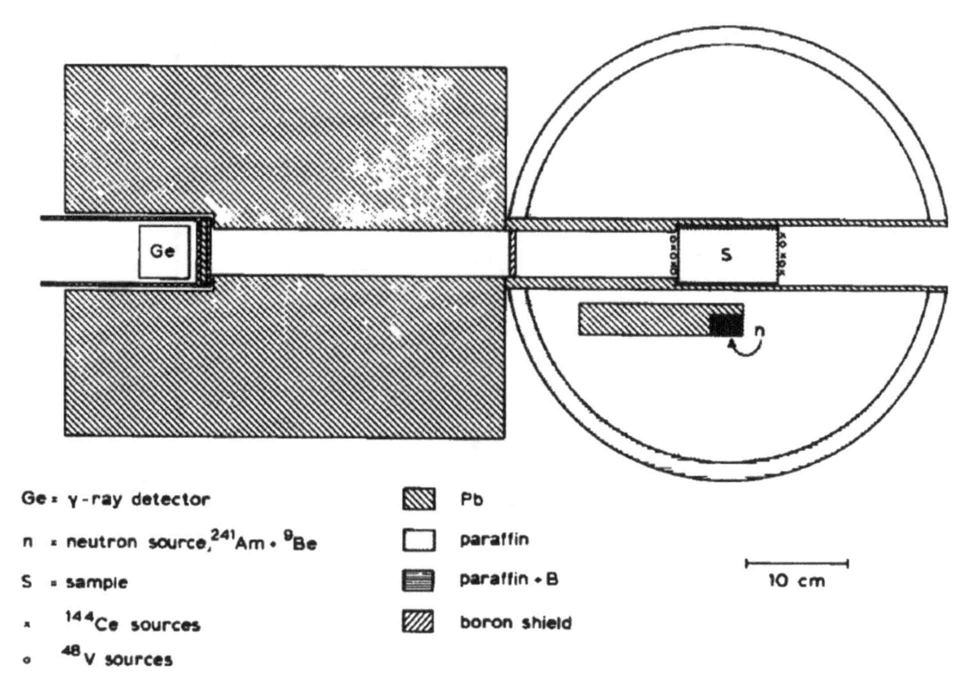

Figure 8.4. Schematic view of the sample and different neutron and γ sources, the paraffin thermalizer and lead shielding of the detector. Sources of ^{48}V and ^{144}Ce were used for calibration. Reproduced with permission from [9]. Copyright 1982 Elsevier.

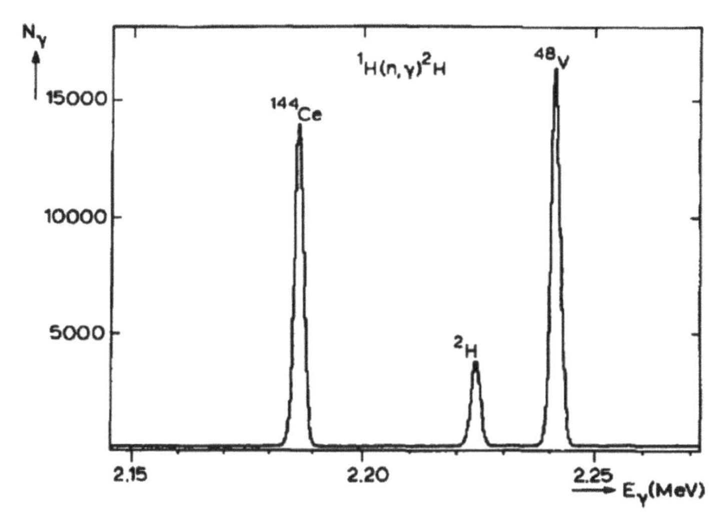

Figure 8.5. The spectrum of p(n, γ)d γ of the n-type intrinsic Ge detector with an energy resolution of 1.75 keV at 1.33 MeV. Sources of ^{48}V and ^{144}Ce were used for calibration. Reproduced with permission from [9]. Copyright 1982 Elsevier.

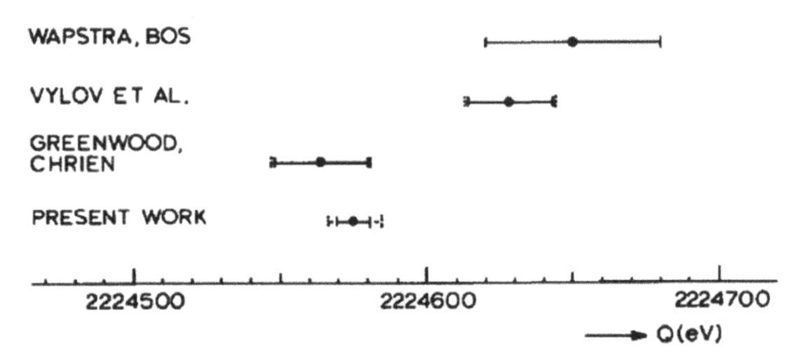

Figure 8.6. The results of [9] in comparison to earlier conflicting results of Wapstra *et al* [11] (compilation of earlier data), Vylov *et al* [10] and Greenwood *et al* [4]. Reproduced with permission from [9]. Copyright 1982 Elsevier.

Bibliography

[1] Audi G, Wapstra A H and Thibault C 2003 *Nucl. Phys.* A **729** 337

[2] Bainbridge K T 1933 *Phys. Rev.* **44** 57

[3] Chadwick J and Goldhaber M 1934 *Nature* **134** 237

[4] Greenwood R C and Chrien R E 1980 *Phys. Rev.* C **21** 498

[5] Helmer R G, Van Assche P H M and van der Leun C 1979 *At. Data Nucl. Data Tables* **24** 39

[6] Mobley R C and Laubenstein R A 1950 *Phys. Rev.* **80** 309

[7] http://physics.nist.gov/cuu/Constants/index.html

[8] Urey H C, Brickwedde F G and Murphy G M 1932 *Phys. Rev.* **40** 1

[9] Van der Leun C and Alderliesten C 1982 *Nucl. Phys.* A **380** 261

[10] Vylov Ta *et al* 1978 *Yad. Fiz.* **28** 1137

[11] Wapstra A H and Bos K 1977 *At. Data Nucl. Data Tables* **19** 175

Chapter 9

The first nuclear reaction with an accelerated beam and the Cockroft–Walton accelerator

In an early paper [2] Cockroft and Walton very clearly outline the necessity of using higher-energy/higher-intensity beams of projectiles for the study of nuclear reactions. They refer to the use of α particles from radioactive sources and the limitations related to this: low intensities (a radium source with an activity equivalent to a decent charged-particle beam would require hundreds of grams of radium!), very limited energy variability and a limitation to α only, etc. They knew that initiating nuclear reactions between charged particles surmounting the Coulomb barrier would require energies of many MeV. But George Gamow *et al* had recently shown that—via quantum-mechanical tunneling—nuclear reactions might be possible at much lower energies with, however, reduced intensities [5, 8].

With this knowledge they discussed in detail the possibilities of creating high voltages for particle acceleration, among them a Tesla-coil device, an ac device and a pulse generator, and concluded that some kind of dc voltage-multiplication scheme using a high-voltage transformer, vacuum-tube rectifiers and capacitors could be suitable. In addition they also discussed methods to produce low-energy high-current ion beams in canal-ray tubes such as those used for mass spectrometry.

In 1932 Cockroft and Walton [3, 4] published the results of their efforts to produce sufficiently high dc voltages to initiate a nuclear reaction. They developed a multiplier circuit using rectifier diodes and capacitors that could withstand high voltages up to 200 kV. Such circuits had been devised elsewhere [6, 7, 10, 11], but Cockroft and Walton adapted them to their special needs. Figure 9.1 explains the principle of the circuit and the vacuum accelerator tube design in several stages connected to different voltage levels from the rectifier. The use of several accelerating gaps has the advantage of a smaller electric field strength per gap and better focusing properties for the ion beam. For a comprehensive review see [1]. Figures 9.2 and 9.3 show a schematic and a photograph of the entire accelerator set-up.

doi:10.1088/978-0-7503-1173-1ch9

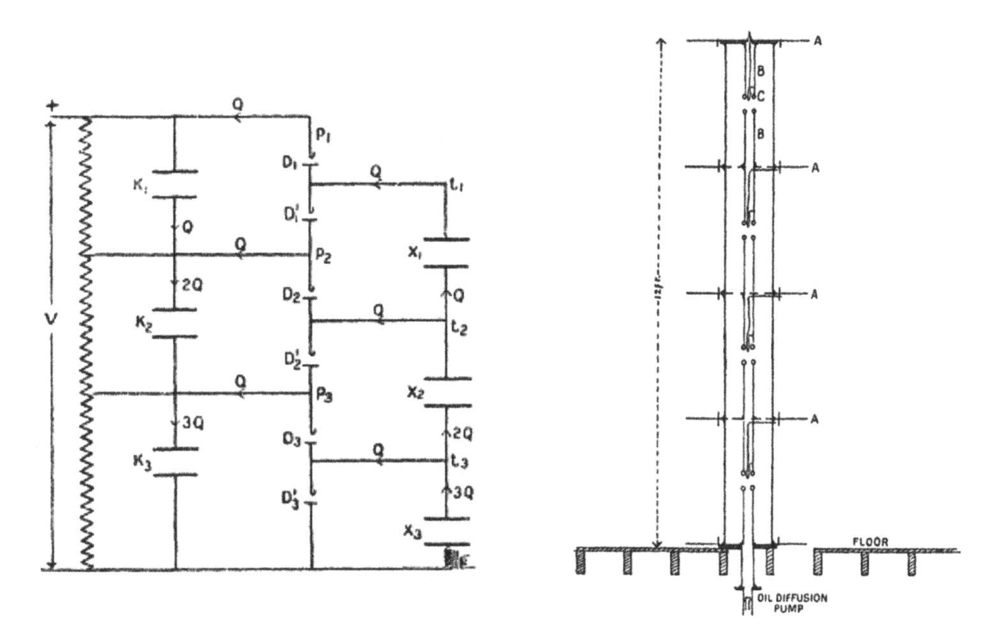

Figure 9.1. Schematic of the voltage-multiplication circuit used in the first 'Cockroft–Walton' accelerator and accelerator tube design. Reproduced from [3].

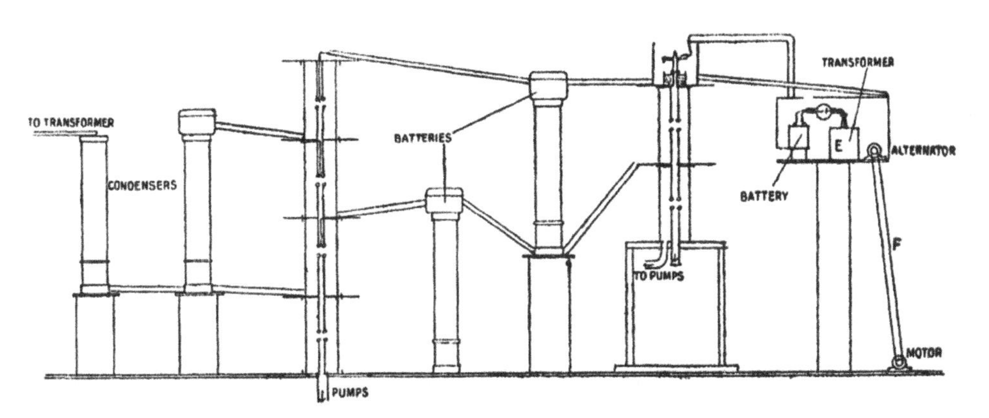

Figure 9.2. Schematic of the accelerator set-up consisting of a three-stage voltage-multiplication circuit connected to the experimental tube in which protons from a discharge canal-ray ion source could be accelerated to 800 keV. Reproduced from [3].

The proton beam of up to 15 μA could be accelerated to about 720 keV and focused only by the accelerating tube arrangement with no extra focusing elements. In the first series of experiments the proton beam, described as a visible 'pencil' exiting the thin mica window into open air, was used to measure the proton range in different gases, see figure 9.4.

In a modified set-up the protons impinged on a Li target under 45° in a chamber shown in figure 9.4. The reaction particles were observed as scintillations on a ZnS

Figure 9.3. Photograph of the accelerator complex. In the box covered by a black cloth the experimenter would sit and count scintillations on a fluorescent screen, e.g. as function of the scattering angle. Reproduced from [3].

screen with a microscope and also in a cloud chamber where, by the thickness of the tracks, they were identified as α particles. Together with a second such microscope set up on the opposite side, frequent coincident emissions of two α particles were registered such that the authors unambiguously concluded that the reaction

$$^{7}\mathrm{Li} + \mathrm{p} \rightarrow 2\alpha + 17.347\ \mathrm{MeV} \tag{9.1}$$

had taken place. This was thus the first nuclear reaction initiated by artificially accelerated projectiles (and also one of the first *coincidence* experiments [4]; the first was that by Bothe [8, 9] for which he earned the Nobel Prize in 1974).

The authors varied the proton energy and saw a typical Coulomb threshold behavior. Then they subjected a remarkably large number of elements to the proton beam and registered the relative number of counts per unit time and beam current (i.e. they measured essentially the cross sections for (p, α) reactions).

The use of accelerated ions started the entire field of nuclear reactions and its rapid development. The development of particle accelerators is a prerequisite not only for the study of nuclear reactions and the elucidation of their mechanisms, but also for nuclear structure studies. The possibility of exciting many levels of thousands of nuclei and to study their formation and decay led to the important *shell model* and different *collective models* describing the rich landscape of nuclear shapes, their motion and their excitations. The progress in our knowledge of these properties is therefore intimately connected with the development of accelerators. The quest for ever higher energies is still with us, but other properties such as higher energy

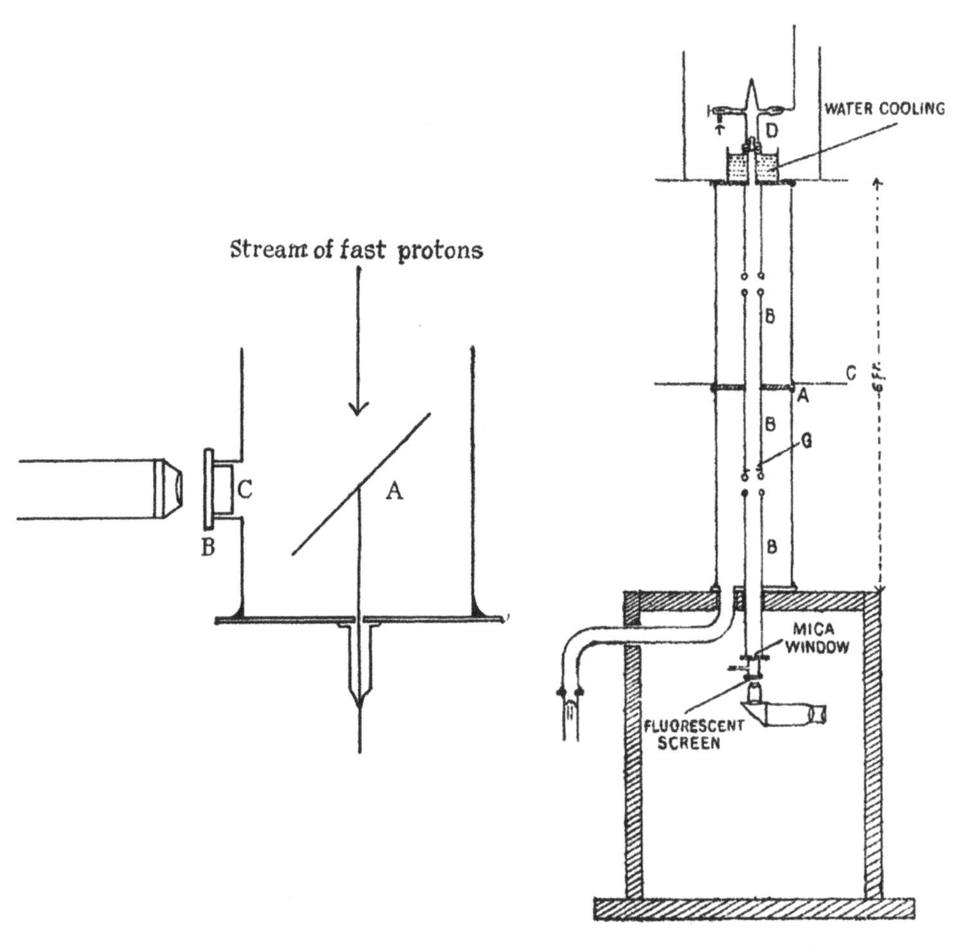

Figure 9.4. Acceleration tube and reaction chamber set-up with a vacuum pump system, a thin mica window allowing protons to exit into air or to impinge on a Li target. In this latter case (left) a microscope was used to count the reaction α particles as scintillations on a fluorescent screen. When using a second scintillator on the opposite side coincident events could be observed. Reproduced from [3, 4].

resolution, ease of changing energies and beam projectile type, and high beam currents for better statistics have also driven the development up to the highest energies, such as that of the LHC with a relative energy (i.e. an energy available in the center-of-mass system of $E = 14\,\text{TeV}$). The Livingston plot in figure 9.5 documents this. Fixed-target experiments—due to relativistic effects—make less and less relative energy available for particle interactions with increasing laboratory energy whereas colliders make the full relative energy of the head-on colliding beams available (with equal masses and the laboratory energies of the two beams twice the laboratory energy). For the study of nuclear reactions, Van de Graaff machines and cyclotrons (in their modern form of (spiral-)sector focusing) are best suited and have become the 'workhorses' of low-energy nuclear physics. The simultaneous development of ion-source technologies, solid-state detectors and magnetic spectrographs

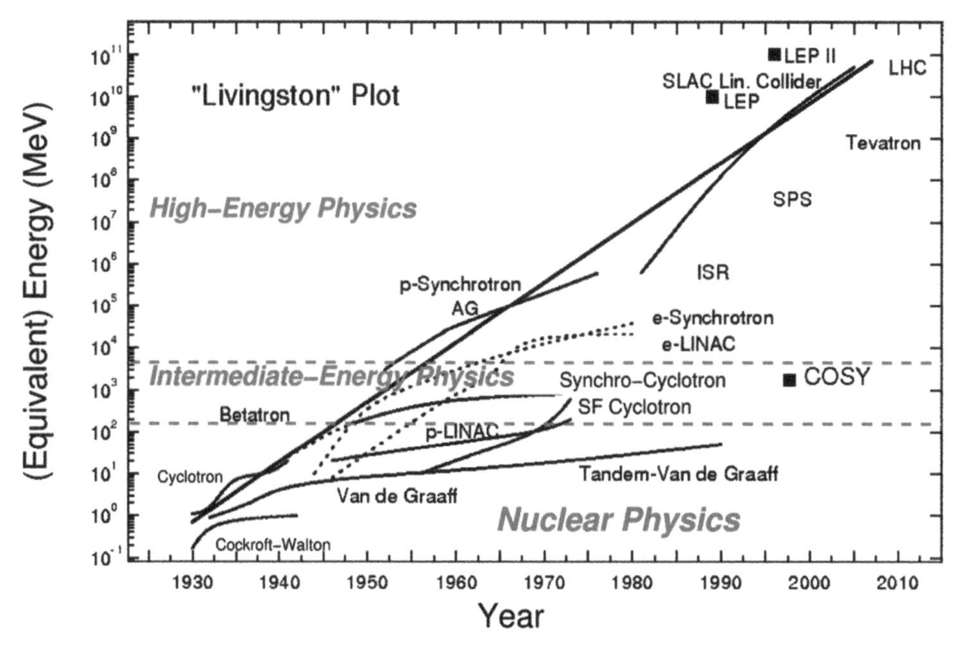

Figure 9.5. 'Livingston plot': a plot of the development of accelerators over the years with a doubling of the 'available' energy approximately every seven years. The approximate energy ranges of (low-energy) nuclear physics proper, of intermediate-energy physics where the overlap and interactions of quarks and nuclei are studied, and high-energy (or particle) physics where all facets of the standard model are investigated, are indicated in the plot.

has made the broad spectrum of different investigations and discoveries in nuclear physics with high-quality data possible. Examples are:

- Fine-structure studies on isobaric analog resonances, possible only through the high energy resolution and fine-tunability of the tandem Van de Graaff accelerator.
- Heavy-ion reaction studies as well as their use in γ spectroscopy through the ease of changing between many heavy-ion species in the beam source, energies and charge states to obtain high beam energies.

Cockroft–Walton machines and cyclotrons are in use as the first part in the chain of accelerators in intermediate- and high-energy installations such as COSY and CERN. Accelerators in the low- and intermediate-energy domain have also found many useful applications in medicine, art history, archeology, etc.

Bibliography

[1] Baldinger E 1959 Kaskadengeneratoren *Nuclear Instrumentation* vol 1 (*Encyclopedia of Physics/Handbuch der Physik* vol 44) ed E Creutz (Heidelberg: Springer)

[2] Cockroft J D and Walton E D S 1930 *Proc. R. Soc.* A **129** 477

[3] Cockroft J D and Walton E D S 1932 *Proc. R. Soc.* A **136** 619

[4] Cockroft J D and Walton E D S 1932 *Proc. R. Soc.* A **137** 229

[5] Gamow G 1929 *Z. Phys.* **52** 510
[6] Greinacher H 1914 *Phys. Z.* **15** 410
[7] Greinacher H 1921 *Z. Phys.* **4** 195
[8] Guerney R W and Condon E U 1929 *Phys. Rev.* **33** 127
[9] Livingston M S 1954 *High-Energy Accelerators* (New York: Interscience)
[10] Schenkel M 1919 *Elektrotechn. Z.* **40** 333
[11] Slepian J 1928 High-voltage direct-current system *US Patent class* 175–363, No. 1,666,473

IOP Publishing

Key Nuclear Reaction Experiments
Discoveries and consequences
Hans Paetz gen. Schieck

Chapter 10

Observation of direct interactions

Compound nuclear (CN) reactions and *direct interactions* (DIs) are two forms of reactions of composite nuclei that are best classified according to their time behavior. They mark the extremes in this classification. The DIs are processes which occur at time scales of the traversal times of projectiles past target nuclei (typical times are 10^{-22} s). At the other end are the CN reactions where projectiles are more or less completely absorbed by the target to form a compound nucleus which will survive a long time during an equilibration process and decay into open channels without memory of the formation process (typical times are 10^{-16} s).

10.1 Elastic scattering and the optical model

Around 1950 it was noticed that the scattering data of intermediate-energy nucleons (in the beginning mainly neutrons) could not be described well in the framework of the CN models. Angular distributions as well as excitation functions showed marked diffraction patterns. Early neutron elastic scattering using 84 MeV neutrons from the Berkeley 184″ cyclotron was reported by Bratenahl *et al* [4], as shown in figure 10.1. Very early proton data were measured by Burkig *et al* [5] using 18.6 MeV protons on several nuclei, among them aluminum. For the description of the elastic scattering data two limiting cases of models were inspired by phenomena in optics, the *opaque* model and the *transparent* model [7, 18]. The p scattering data on Al were fitted to two different optical model theories by Lelevier *et al* [13]. For the one that gave a better fit to the diffraction pattern the authors coined for the first time the name optical model. A simple complex square-well potential $V(r) = -(V_0 + \mathrm{i}W_0)$; $r < R_0$ with a fixed potential radius of $R_0 = 1.42A^{1/3}$ fm was assumed. A best fit was obtained with $V_0 = 45$ MeV and $W_0 = 20$ MeV, values in agreement with neutron scattering. A picture of the first 'optical model' fit is shown

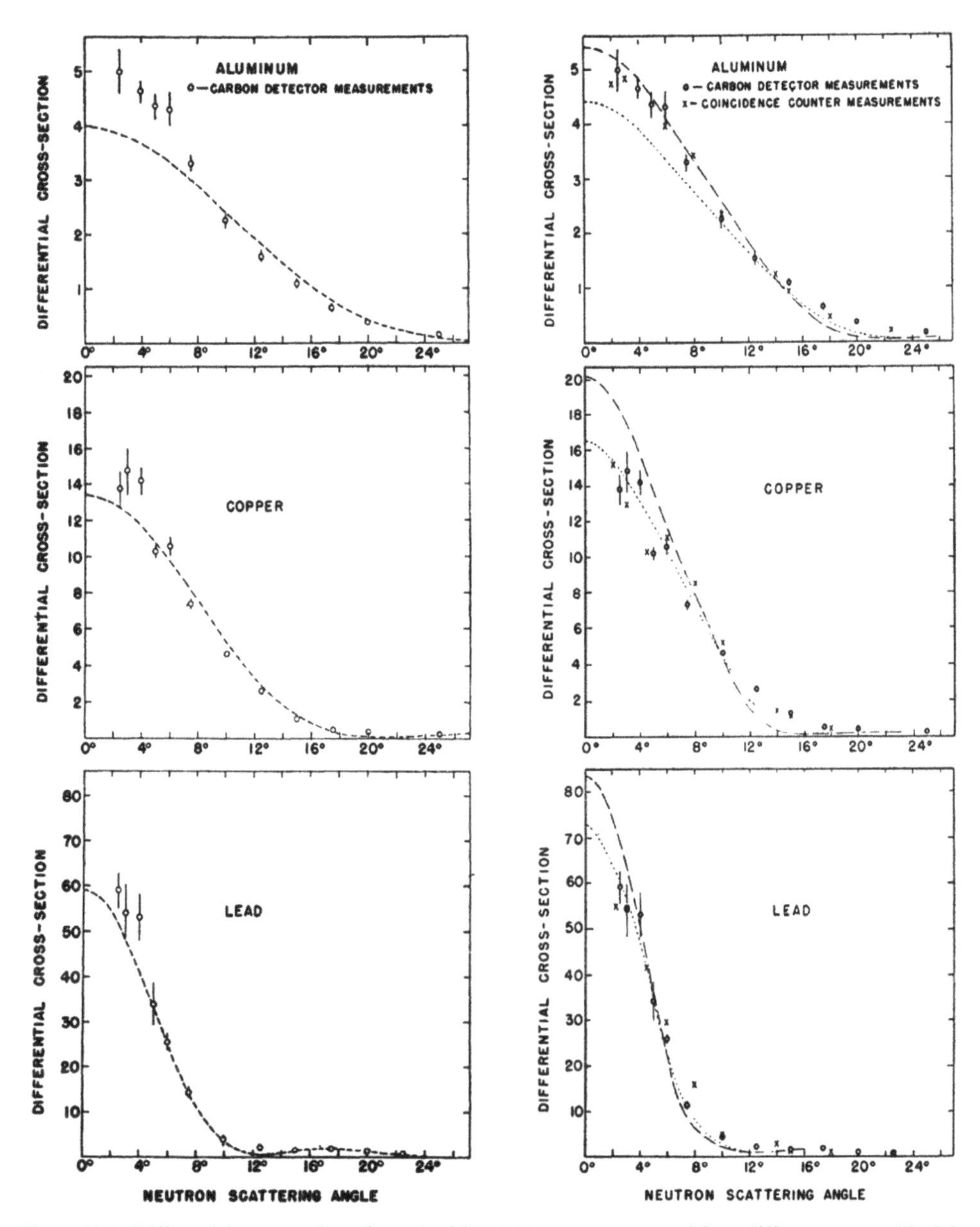

Figure 10.1. Differential cross sections (in b/sr) of 84 MeV neutrons scattered from different targets. The left-hand panels show the predictions for an 'opaque-nuclear' model where the dashed (dotted) curves are the predicted patterns of σ_s' and $\sigma_s' - \sigma_2$. The right-hand panels show the predictions of a 'transparent' model. Reproduced with permission from [4]. Copyright 1950 American Physical Society.

in figure 10.2. Pasternack *et al* [15] had also obtained good fits to neutron data with a transparent model with *absorption* in the WKB approximation. The basic features of the 'standard optical model', developed after the key experiments, will be briefly described here and only a few standard references are given: [10, 11, 14].

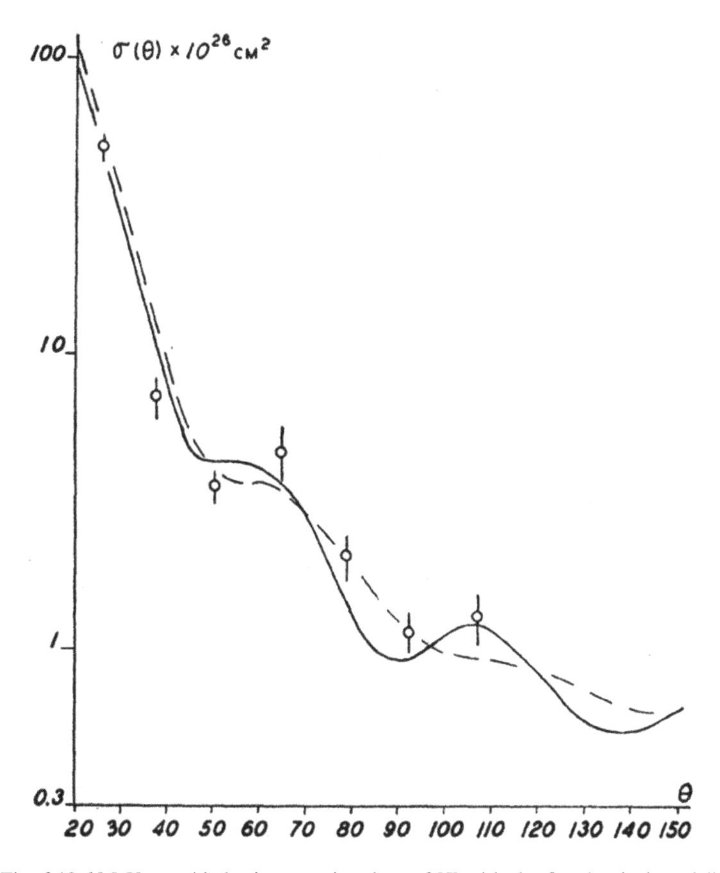

Figure 10.2. Fit of 18.6 MeV p + Al elastic scattering data of [5] with the first 'optical model'. Reproduced with permission from [13]. Copyright 1950 American Physical Society.

The radial Schrödinger equation for protons (spin $s = 1/2$) reads:

$$\left[\frac{\mathrm{d}^2}{\mathrm{d}r^2} + k^2 - \frac{\ell(\ell+1)}{r^2} + Vf(r) + \mathrm{i}Wg(r) - V_\mathrm{C}(r) + \left(V_{\mathrm{s.o.}} + \mathrm{i}W_{\mathrm{s.o.}}\right)h(r) \cdot \begin{Bmatrix} \ell \\ -(\ell+1) \end{Bmatrix}\right] u_{\ell j}^{(\pm)}(kr) = 0, \tag{10.1}$$

where V and W are the real and imaginary parts of the central potential, V_C is the Coulomb potential, and $V_{\mathrm{s.o.}}$ and $W_{\mathrm{s.o.}}$ are the real and imaginary parts of the spin–orbit potential, particularly important for the description of polarization observables. For neutrons the Coulomb term vanishes. The solution of this equation is part of the total scattering function

$$\Psi = \frac{1}{kr}\sum_{\ell j \lambda} \mathrm{i}^\ell \left[4\pi(2\ell+1)\right]^{1/2} (\ell 0 s\mu \,|\, jm)(\ell\lambda s\nu \,|\, jm) u_{\ell j}^{(\pm)}(kr) Y_\ell^\lambda(\theta, \phi) \chi_s^\mu \mathrm{e}^{\mathrm{i}\sigma_\ell} \tag{10.2}$$

with

$$u_{\ell j} \rightarrow_{r\to\infty} \frac{1}{2\mathrm{i}}\left[\mathrm{e}^{-\mathrm{i}(kr-\eta\ \ln 2kr-\ell\pi/2)} - \mathrm{e}^{2\mathrm{i}\delta_{\ell j}}\mathrm{e}^{\mathrm{i}\left(kr-\eta\ \ln 2kr-\ell\pi/2+2\sigma_\ell\right)}\right]. \tag{10.3}$$

$\delta_{\ell j}$ are the complex nuclear scattering phases, $\sigma_\ell = \arg\Gamma(1+\ell+\mathrm{i}\eta)$ the Coulomb scattering phases, $\eta_\ell^j = \mathrm{e}^{2\mathrm{i}\delta_{\ell j}}$ the 'reflection coefficients', $\eta = Z_1Z_2e^2/\hbar v$ is the Coulomb (or Sommerfeld) parameter (see 2.39) and $k = \sqrt{2\mu E_{\mathrm{kin}}^{\mathrm{c.m.}}/\hbar^2}$ the entrance-channel wavenumber.

The potential form factor $f(r)$ is defined in analogy to the shape of the usual nuclear density or potential distributions that are used in the classical shell model (the *Woods–Saxon form*). The absorption occurs predominantly at the nuclear surface. Thus, as the form for $g(r)$ at low energies one chooses the derivative of the Woods–Saxon form factor and for the spin–orbit term h the Thomas form $g(r)/r$. At higher energies more absorption in the nuclear volume is plausible, which is taken into account by a gradual transition from the surface absorption to volume absorption. The best sets of parameters have been obtained by fits with χ^2 minimization to a large number of data sets of cross sections as well as analyzing powers. The latter are important for fixing the $(\ell\cdot\mathbf{s})$ potential and removing typical ambiguities in the potential parameters. A few of these parameter sets have become standards for the optical model. For nucleon scattering the parametrization most used is that of Greenlees and Becchetti [2], especially because they provide a global set of parameters (i.e. valid over a large region of the periodic table). However, in special cases, e.g. near doubly magic nuclei, this set is not as good as a single fit. It is interesting that the depth of the real potential corresponds closely to that of the shell model potential, similarly for the $(\ell\cdot\mathbf{s})$ term. For light projectiles consisting of A nucleons (deuterons, α particles, etc) one has potential depths that are A times the nucleon potential depths. For heavy-ion scattering there are quite different approaches, some with very shallow potentials. Figure 10.3 shows the form factors and the behavior of imaginary potentials with energy. Figure 10.4 shows the result of a global fit to 14.5 MeV proton scattering data over a large mass range [2]. The slow variation of the diffraction-like pattern points to slow changes of geometrical potential parameters such as the potential radius. Figure 10.5 shows the result of an optical model calculation of elastic proton scattering on ^{90}Zr as a function of incident energy and angle [17] with parameters from [2]. Slow variations with mass number and with energy (with wide single-particle resonance structures) and diffraction-like angular structures are indicative of a DI process. The energetic widths of these structures as well as their strength function dependence on A suggest single-particle interactions with the nuclei as a whole ('giant' or 'size' resonances), see e.g. [19, 20]. The extremely deep backward minima shown in figure 10.5 make CN contributions (as Ericson fluctuations) visible, see [3]. It can only be mentioned in passing that steps to found the optical potentials on more microscopic grounds have been undertaken by creating *folding potentials*. In these the potential of one nucleon of the projectile with the target nucleus is folded with the nucleon density of the projectile nucleus and vice versa, or the same for both nuclei (*double* folding potentials).

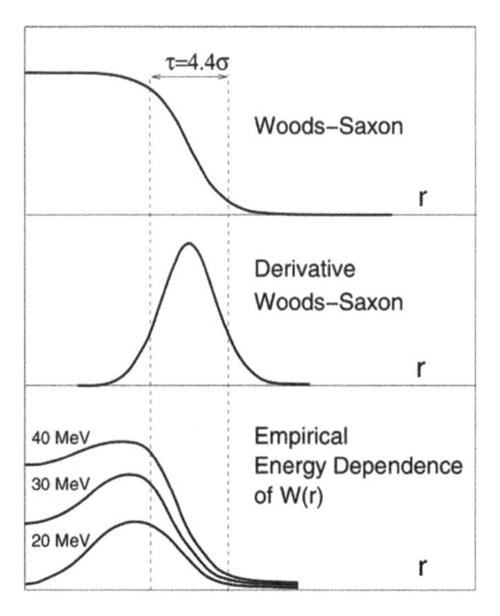

Figure 10.3. Form factors of the optical model. (Top) Woods–Saxon form of the real part f. (Middle) Derivative Woods–Saxon form $g = f'$ of the imaginary part. (Bottom) Sliding-transition form of the surface-to-volume imaginary part as function of energy.

10.2 Direct (rearrangement) reactions

A multitude of these reactions may be classified:

- Reactions without change of the mass number.
  - Elastic potential scattering (see above; description by the optical model).
  - Direct inelastic scattering ((p,p$'\gamma$), (α,α'), ...). It leads preferentially to collective nuclear excitations (such as rotation, vibration, etc).
  - Quasi-elastic (charge-exchange) processes ((p,n), (n,p), (^{3}He,t), (^{14}N,^{14}C), ...). These lead, e.g., to isobaric-analog states of the target nucleus.
- Reactions with change of the mass number.
  - Pickup reactions (one-nucleon transfer: (p,d), (d,^{3}He), (d,t), ...; few or multi-nucleon transfer: (p,α), (d,^{6}Li), ...).
  - Stripping reactions (one-nucleon transfer: (d,p), (d,n), (^{3}He,d), ...; few- or multi-nucleon transfer: (^{6}Li,d), (α,p), (^{3}He,p), ...).
  - Knockout reactions ((p,α), (p,p$'$), ...).
  - Direct breakup processes such as knockout with few-particle exit channels ((p,pp), (α,2α), ...).
  - Induced fission is a special case of a rearrangement reaction resulting in larger debris.
  - Processes of higher order (multi-step processes via excited intermediate states, coupled channels).

Here only the simplest case of the *stripping reaction* will be discussed. The many details of DI processes are the subject of a large number of books, see e.g. [1, 8, 9, 16].

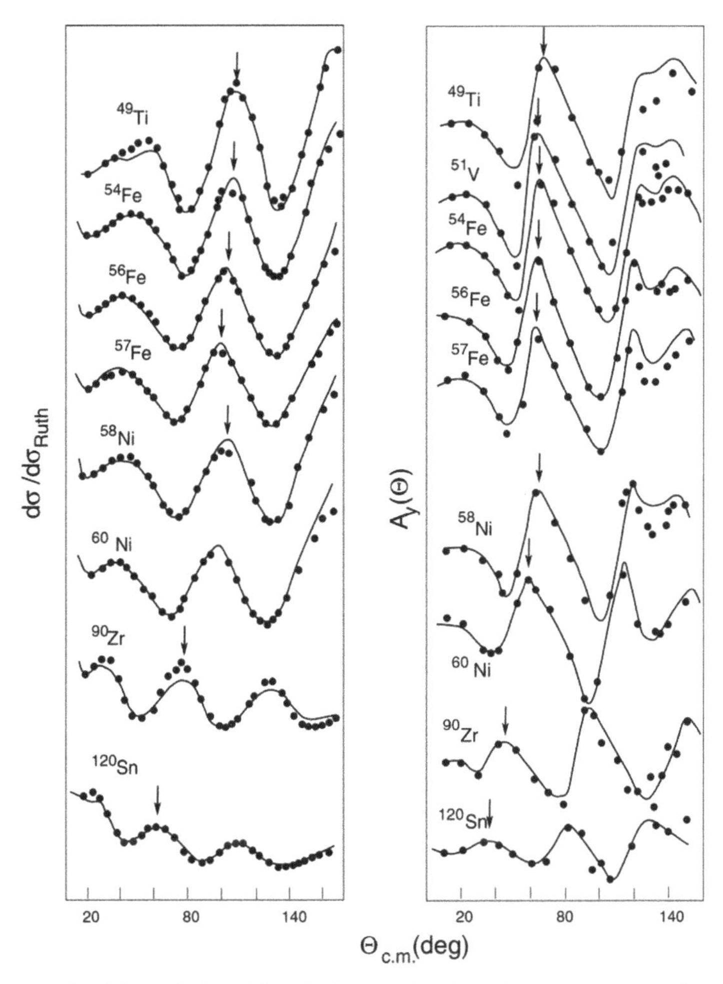

Figure 10.4. Global fit of the optical model to elastic scattering data of 14.5 MeV protons for a large nuclear mass range. The figure shows (with arbitrary scales) cross sections normalized to the Rutherford cross sections (i.e. they should be 1 at 0°), the analyzing powers are 0 at 0°. The arrows indicate the systematic variation of characteristic diffraction maxima with the target mass. Adapted with permission from [2]. Copyright 1969 American Physical Society.

Standard computer codes such as DWUCKn (distorted wave code) and CHUCKn (coupled channels distorted wave code) [12] are available. For induced nuclear fission see chapter 14.

10.3 Stripping reactions

Already a semi-classical ansatz provides a qualitative picture of the angular distributions of stripping reactions. It explains the expected behavior with the assumption of a rapid process that is localized at the nuclear surface and is non-equilibrated. The wave-number vector of the incoming deuterons is $\vec{k}_d$, those of the transferred nucleon and of the outgoing nucleons are $\vec{k}_n$ and $\vec{k}_p$, respectively.

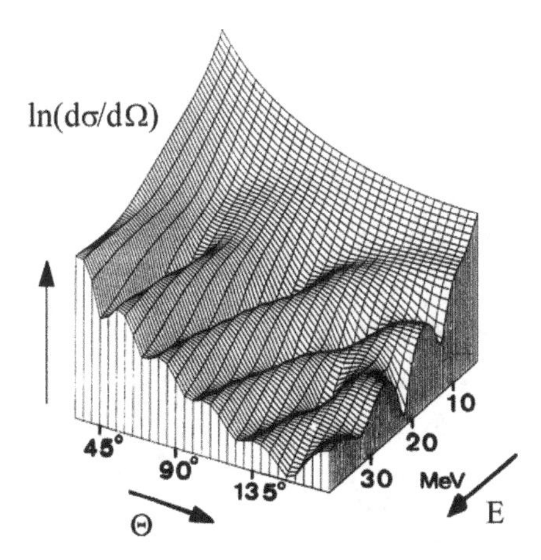

Figure 10.5. Angular and energy dependence of the cross section of elastic proton scattering from ^{90}Zr calculated with standard Greenlees–Becchetti global parameters of the optical model [2]. The interference structure of the angular distributions may be interpreted as 'resonant' (single-particle) structures of the excitation function with widths typical for fast (i.e. direct) processes. They are also analogous to diffraction structures in classical optics. Reproduced with permission from [17]. Copyright 1975 Elsevier.

They form a momentum diagram, from which the connection between a preferred scattering angle θ and the transferred momentum and also the angular momentum can be deduced:

$$p_n R = \hbar k_n R = \hbar \ell_n. \tag{10.4}$$

For small θ

$$\theta_0 \approx \frac{k_n}{k_d} = \frac{\ell_n}{k_d R}. \tag{10.5}$$

Because of the quantization of ℓ there are discrete values of θ increasing with ℓ. This qualitative picture is not changed when calculating the angular distributions quantum-mechanically. As an example, for the reaction $^{52}\mathrm{Cr}(\vec{\mathrm{d}},\mathrm{p})^{53}\mathrm{Cr}$ the angles of the stripping maximum calculated in different ways are

	θ^{DWBA}	θ^{PWBA}	$\theta_{\mathrm{s.c.}}$
$\ell = 0$	0^0	0^0	0^0
$\ell = 1$	18^0	13^0	13^0
$\ell = 2$	34^0	19^0	26^0
$\ell = 3$	49^0	30^0	39^0
$\ell = 4$	64^0	40^0	52^0

(10.6)

The measured angular distributions of the cross sections show—in addition to diffraction structures—marked stripping maxima, the angular position of which often allows the determination of the angular momentum of the transferred nucleon. An example is shown in figure 10.6. Historically this feature was (and still is)

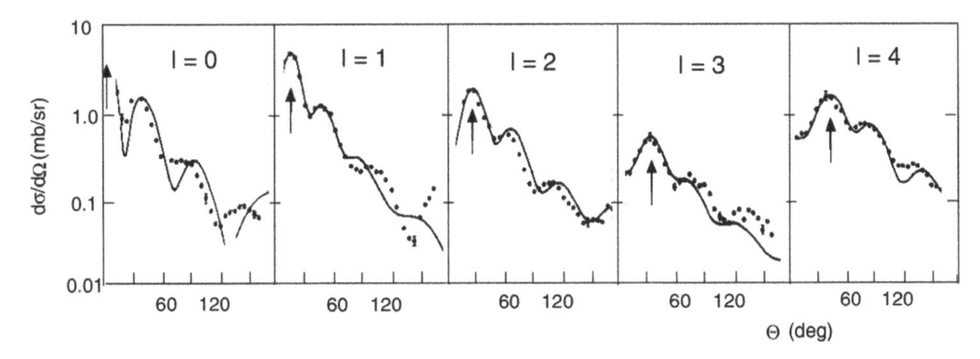

Figure 10.6. Characteristic and systematic features of the stripping maximum as a function of the transferred orbital angular momentum ℓ. The arrows indicate the change of the angular position of the stripping peak with ℓ. Adapted with permission from [6]. Copyright 1967 American Physical Society.

important for the assignment of the final nuclear states of a reaction to orbitals in the one-particle shell model. The energy relations in stripping reactions are such that the transferred nucleon near magic shells is preferentially inserted into low-lying shell-model states. Because of the spin–orbit splitting the complete assignment also requires that the total angular momentum j of the transferred nucleons is known. A good method is the measurement of the analyzing power of the stripping reaction, i.e. the use of polarized projectiles. In many cases the distinction between the two possibilities for j can be made just from the sign of the analyzing power alone. For an intuitive description of this fact there exists again a simple semi-classical model (Newns). It is assumed that the interaction happens at the nuclear surface and that we have a relatively strong absorption in the nuclear matter. Thus the front side of the nucleus directed towards the projectile contributes more strongly to the reaction than the backside of the target nucleus. In the front part the orbital angular-momentum vector points upward perpendicularly to the reaction plane whereas in the back half the orbital angular momentum points down. If the incident deuteron is polarized up or down perpendicularly to the scattering plane—under the assumption of the existence of a spin–orbit force—the transferred nucleon in the scattering to the left ends preferentially in a state with $j = \ell + 1/2$ for the up case, in the down case with $j = \ell - 1/2$. The measured analyzing power

$$A_y = \frac{1}{P_d}\frac{N^{\text{up}} - N^{\text{down}}}{N^{\text{up}} + N^{\text{down}}}, \tag{10.7}$$

will show opposite signs for the two cases [21]. Figure 10.7 shows a good example of the stripping reaction $^{40}\text{Ca}(\vec{\text{d}},\text{p})^{41}\text{Ca}$. This behavior has been confirmed for many examples, not only for stripping reactions. If one wants to know the degree to which the transition considered is a single-particle transition the *spectroscopic factor* has to be determined. For that—at least approximately—quantitative theories are required.

Because the single-particle strength is often strongly fractionated by the residual interaction, i.e. spread out over many states in a range of energies, spectroscopic investigations on many final nuclear states are necessary. Often these states are close

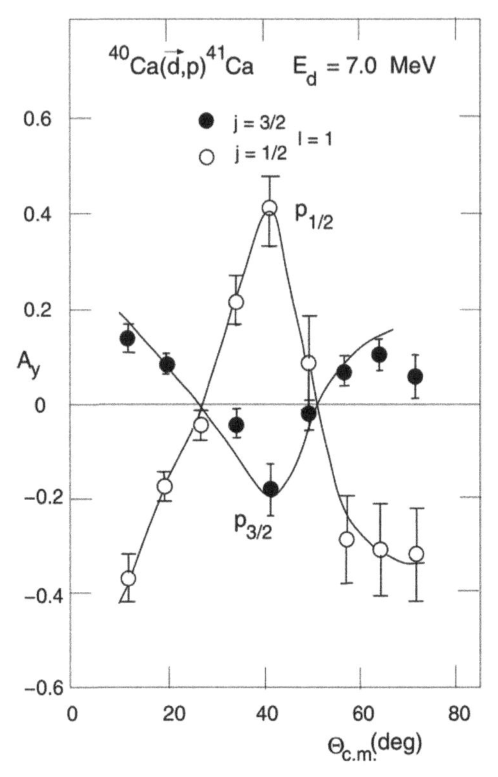

Figure 10.7. Sensitivity (sign!) of the analyzing power of the stripping reaction $^{40}Ca(\vec{d},p)^{41}Ca$ to the total angular momentum $j = \ell \pm 1/2$ of the transferred nucleon. Adapted with permission from [21]. Copyright 1968 Elsevier.

together and a high detector resolution is necessary to obtain a complete picture. Particularly useful tools for this purpose are magnetic spectrographs with high resolution at tandem Van de Graaff accelerators, also with polarized particle beams.

10.4 The Born approximation

Here one uses the first Born approximation, i.e. the first term of the Born series. Starting from Fermi's golden rule of perturbation theory, which predicts for the differential cross section

$$\frac{\mathrm{d}\sigma}{\mathrm{d}\Omega} = \frac{\left(2I_b + 1\right)\left(2I_\mathrm{B} + 1\right)}{2\pi^2\hbar^4}\mu_i\mu_f\frac{k_f}{k_i}\left|T_{if}\right|^2, \tag{10.8}$$

one has to make assumptions about the transition matrix element.

In the *plane wave Born approximation* (PWBA) (also Butler theory) for the incoming and outgoing waves plane waves are used. Since the radial wave functions are Bessel functions one finds a simple diffraction pattern for the cross section

$$\frac{\mathrm{d}\sigma}{\mathrm{d}\Omega} \propto \left[j_\ell(kR)\right]^2. \tag{10.9}$$

.

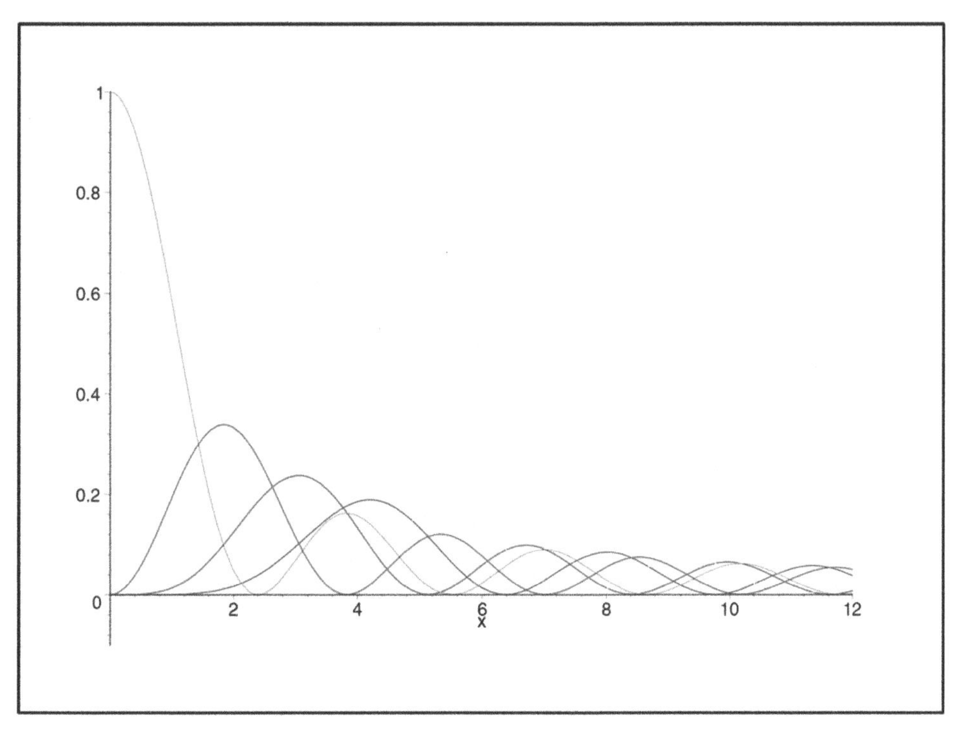

Figure 10.8. The behavior of the squares of the lowest-order spherical Bessel functions $j_\ell(kr)$ as functions of $x = kr$.

For illustration figure 10.8 shows the few lowest-order spherical Bessel functions squared. The angle dependence of the stripping maximum is contained only in the momentum relation $k^2 = k_{\text{in}}^2 + k_{\text{out}}^2 - 2k_{\text{in}}k_{\text{out}} \cos\theta$. Only in a few simple cases is the angular distribution near the maximum satisfactorily described by PWBA. It also makes no statements about polarization observables and contains no information about nuclear structure.

Better results, at least for forward angles, are obtained with the *distorted wave Born approximation* (DWBA). It was formulated with a number of additional and far-reaching assumptions:

- In the entrance and exit channels *distorted* waves are used, i.e. the wave functions are the solutions obtained from fits of the optical model to the elastic scattering data in each pertaining channel at the proper channel energy. For example, for the description of the reaction A(d,p)B the optical model wave functions from the fit to the data of the scattering A(d,d)A as well as of B(p,p)B are needed. Thus—still in the first Born approximation—the diffraction and absorption of the incoming and outgoing waves in the nuclear (and eventually the Coulomb) field, as well as the effect of the $(\ell \cdot \mathbf{s})$ potential,

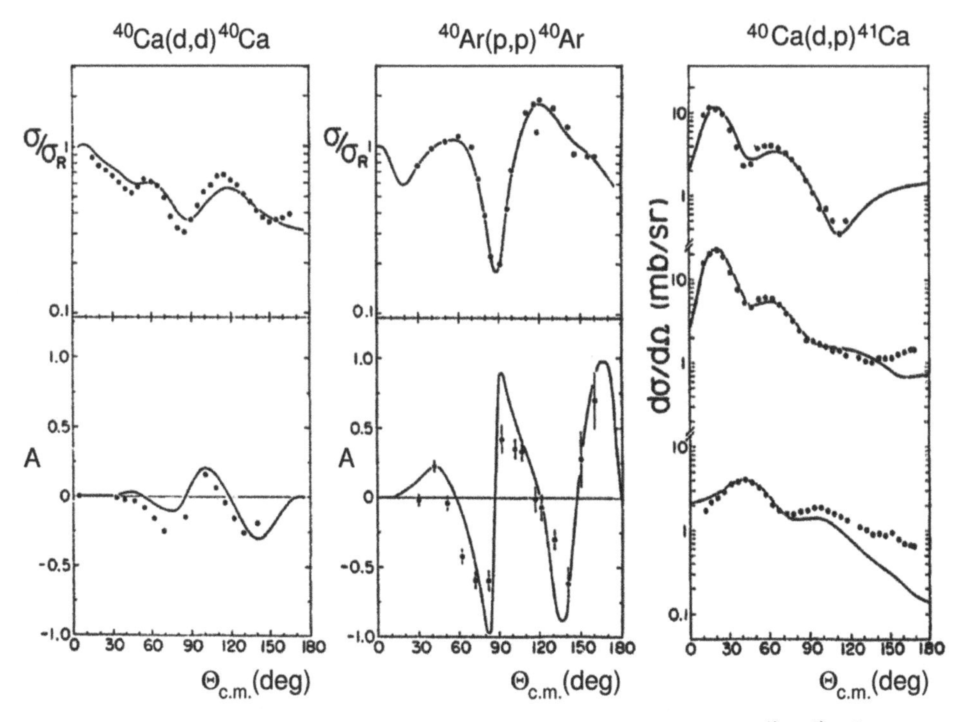

Figure 10.9. Related optical model and DWBA calculations for the stripping reaction $^{40}Ca(\vec{d},p)^{41}Ca$. Shown are the optical model fits to the entrance and exit elastic scattering channel data (left and center) leading to distorted wave functions used in the DWBA calculations (right). Lacking a suitable target ^{41}Ca the neighboring ^{40}Ar has been used relying on the weak mass dependence of the optical model parameters. Reproduced with permission from [21]. Copyright 1968 Elsevier.

are also taken into account to describe the polarization data. An example is shown in figure 10.9.

- The nuclear initial and final states are shell model states.
- The finite range of the nuclear forces is taken care of by a *finite-range* or even *zero-range* approximation.
- The T matrix is expanded into partial waves belonging to fixed angular-momentum transfer.
- The transfer matrix is factorized into a nuclear-structure-dependent and into a kinematical part.

Thus the cross section reads:

$$\frac{\mathrm{d}\sigma}{\mathrm{d}\Omega} = \frac{\mu_a \mu_b}{\pi \hbar^4}\left(\frac{m_B}{m_A}\right)^4 \frac{2J_B + 1}{\left(2J_A + 1\right)\left(2s_a + 1\right)} \frac{1}{k_a k_b} \sum_{\ell sj}\left[\left|A_{\ell sj}\right|^2 \sum_m \left|\beta_{sj}^{\ell m}\right|^2\right] \tag{10.10}$$

The experimental cross section is a product of a fit parameter, the spectroscopic factor $S_{\ell j}$ and a theoretical cross section calculated in the framework of the DWBA with the assumption of single-particle states:

$$\left(\frac{d\sigma}{d\Omega}\right)^{\ell j}_{\mathrm{exp}} = S_{\ell j}\left(\frac{d\sigma}{d\Omega}\right)^{\ell j}_{\mathrm{DWBA}}. \tag{10.11}$$

In a stripping process the spectroscopic factor is the square of the amplitude of a fragment of a single-particle state of the final nucleus. Because of this fractionization (which in reality is caused by the residual interaction of the many other nucleons) into many states with equal quantum numbers the *strengths* of all these states have to be summed up. If a complete collection from all these states is possible one obtains the total strength, which can also be calculated because the number of nucleons N in a subshell is known. Therefore sum rules can be applied, e.g. for single-particle stripping $\sum S_{\ell j} = (2J + 1)$. Mathematically the spectroscopic factor is the overlap integral between the anti-symmetrized k-particle final nuclear state $\Psi_A(i)$, into which the nucleon is inserted, and the single-particle configuration of the anti-symmetrized $(k - 1)$-particles target, the nuclear ground state (core) and the single-particle wave function of the transferred kth particle $\Psi(j)$. It thus gives the probability with which a certain state is present in this configuration. When averaging over the strength distribution of all states that are fractions of one single-particle state (e.g. while assuming a Breit–Wigner distribution function) the position of the average energy provides the energy of the single-particle state, whereas the width of the distribution is a measure of its lifetime, the (*spreading width*) $\Gamma^{\downarrow}$. It measures the decay of the single-particle state into the real nuclear states, split and spread out by the residual interaction, and thus its strength.

Bibliography

[1] Austern N 1970 *Direct Nuclear Reaction Theory* (New York: Wiley)

[2] Becchetti Jr F D and Greenlees G W 1969 *Phys. Rev.* **182** 1190

[3] Berg G, Kühn W, Paetz gen. Schieck H, Schulte K and von Brentano P 1975 *Nucl. Phys.* A **254** 169

[4] Bratenahl A, Fernbach A, Hildebrand R H, Leith C E and Moyer B J 1950 *Phys. Rev.* **77** 597

[5] Burkig J W and Wright B T 1951 *Phys. Rev.* **82** 451

[6] von Ehrenstein D and Schiffer J P 1967 *Phys. Rev.* **164** 1374

[7] Fernbach S, Serber R and Taylor T B 1949 *Phys. Rev.* **75** 1352

[8] Glendenning N K 1963 Nuclear stripping reactions *Ann. Rev. Nucl. Sci.* **13** 191

[9] Glendenning N K 1983 *Direct Nuclear Reactions* (New York: Academic)

[10] Hodgson P E 1963 *The Optical Model of Elastic Scattering* (Oxford: Oxford University Press)

[11] Hodgson P E 1967 The optical model of the nucleon–nucleus interaction *Ann. Rev. Nucl. Sci.* **17** 1

[12] Kunz P D, University of Colorado, private communication

[13] Le Levier R E and Saxon D S 1952 *Phys. Rev.* **87** 40

[14] Marmier P and Sheldon E 1970 *Physics of Nuclei and Particles* vol 2 (New York: Academic)
[15] Pasternack S and Snyder H S 1950 *Phys. Rev.* **80** 921
[16] Satchler G R 1983 *Direct Nuclear Reactions* (Oxford: Oxford University Press)
[17] Schulte K, Berg G, von Brentano P and Paetz gen. Schieck H 1975 *Nucl. Phys.* A **241** 272
[18] Serber R 1947 *Phys. Rev.* **72** 1114
[19] Vogt E 1968 *The Statistical Theory of Nuclear Reactions* (*Advances in Nuclear Physics* vol 1) (New York: Plenum) p 261
[20] Vogt E 1972 *Rev. Mod. Phys.* **34** 723
[21] Yule T J and Haeberli W 1968 *Nucl. Phys.* A **117** 1

IOP Publishing

Key Nuclear Reaction Experiments
Discoveries and consequences
Hans Paetz gen. Schieck

Chapter 11

Resonances and compound reactions

11.1 Generalities

Resonances are a very general phenomenon in nature and therefore in all of physics, see also chapter 6. In classical physics they appear when a system capable of oscillations is excited with one or more of its eigenfrequencies, which—depending on the degree of damping—may lead to large oscillation amplitudes of the system. Nuclei are no exception. When tuning the system (changing the exciting frequency) these amplitudes pass through a resonance curve of *Lorentz form*. In particle physics most of the many known 'particles' actually appear as resonances, i.e. as quantum states in the continuum, which decay with characteristic widths (or, equivalently, lifetimes).

Resonances can be discussed in the energy picture where, as a function of energy, excursions of Lorentz form (Breit–Wigner form) with a width Γ appear, and also in the complementary time picture where they appear as quantum states in the continuum, i.e. as states which decay with finite lifetime τ. Between them there is the relation

$$\Gamma = \hbar/\tau. \tag{11.1}$$

In nuclear physics resonances appear in the continuum (i.e. in scattering situations, at positive total energy) when the projectile energy in the center-of-mass system plus the Q-value of the reaction just equals the excitation energy of a nuclear state.

The excitation functions of observables such as the cross section show characteristic excursions from the smooth background when varying the incident energy. The background may be due to a *direct-reaction* contribution from Coulomb or shape-elastic scattering or—in a region of high-level density—may be the energy-averaged cross section of unresolved overlapping compound resonances; in this case resonant excursions would be due to doorway mechanisms. Likewise the scattering phases and scattering amplitudes change in characteristic ways over comparatively small energy intervals. Nuclei may be excited into collective modes such as rotations

doi:10.1088/978-0-7503-1173-1ch11

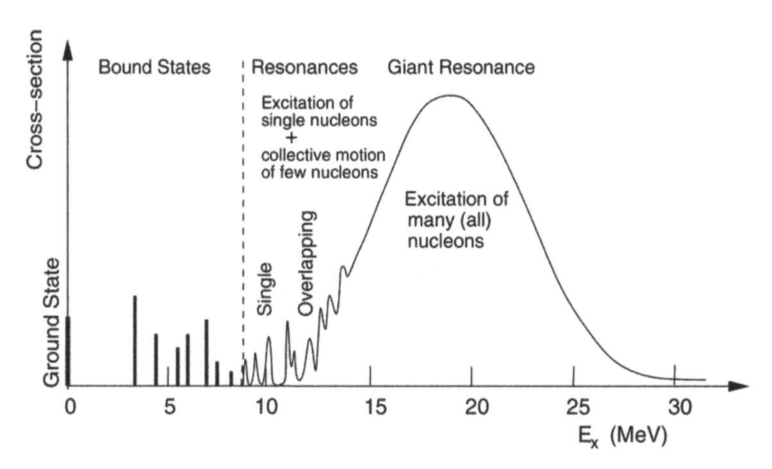

Figure 11.1. The excitation of single resonances, overlapping resonances (with and without Ericson fluctuations) and giant resonances as functions of the energy in the continuum region above the bound-state energy.

and/or vibrations of a part of the nucleons. At still higher energies new phenomena with high cross sections in charged-particle, neutron, γ and π induced reactions appear involving up to all nucleons of a nucleus, the *giant resonances.* Figure 11.1 shows the phenomena schematically in different energy regions.

11.2 Theoretical shape of the cross sections

A model assumption for resonances is—in contrast to direct processes—that the system goes via an intermediate state from entrance into the exit channel. For this case perturbation theory gives the following form of the transition matrix element

$$\langle \Psi_{\text{out}} | H_{\text{int}} | \Psi_{\text{in}} \rangle = \frac{\text{const}}{E - \tilde{E}_{\text{R}}}. \tag{11.2}$$

E_{R} is the energy of the nuclear eigenstate. However, since it is a state in the continuum it is not stationary but one which decays in time. Such states are best described by giving them a complex eigen-energy:

$$\tilde{E}_{\text{R}} = E_{\text{R}} + \text{i}\Gamma/2. \tag{11.3}$$

The interpretation of the imaginary part is: the time development of a state has the form $\text{e}^{\text{i}\tilde{E}t/\hbar}$, on the other hand the state decays with a lifetime τ, whence

$$1/\tau = \text{Im}(\tilde{E}) = \Gamma/2. \tag{11.4}$$

The resonance amplitude thus has the form:

$$g(E) = \frac{F(E)}{E - E_{\text{R}} + \text{i}\Gamma/2}. \tag{11.5}$$

The meaning of $F(E)$ has to be determined. In the sense of Bohr's independence hypothesis, the formation and decay of a resonance are independent (i.e. decoupled). Therefore, one writes the amplitude as the product of the probability amplitude for its formation and its probability of decaying into the considered exit channel. In general, for one formation channel (the entrance channel c) there will be several exit channels c'.

The width Γ of the Breit–Wigner function is inversely proportional to the formation probability P and is the integral over the cross section in the energy range of the resonance:

$$P = \int \frac{\sigma_{aA} v_{aA}}{V} \cdot \frac{V p_{aA}^2 \mathrm{d}p_{aA}}{2\pi^2 \hbar^3} = \int \frac{\sigma_{aA} k_{\mathrm{in}}^2 \mathrm{d}E}{2\pi^2 \hbar} \approx \frac{k_{\mathrm{in}}^2 F(E_{\mathrm{R}})}{2\pi^2 \hbar} \int \frac{\mathrm{d}E}{(E - E_{\mathrm{R}})^2 + \Gamma^2/4}$$
$$= \frac{k_{\mathrm{in}}^2 F(E_{\mathrm{R}})}{\pi \hbar \Gamma}. \quad (11.6)$$

In equilibrium this is equal to the probability that the resonance re-decays into the entrance channel (the purely elastic case). A measure for this is the partial width Γ_{aA} formed similarly to Γ, thus:

$$\Gamma_{aA}/\hbar = \frac{k_{\mathrm{in}}^2 F(E_{\mathrm{R}})}{\pi \hbar \Gamma}, \quad (11.7)$$

$$F(E_{\mathrm{R}}) = \frac{\pi}{k_{aA}^2} \Gamma_{aA} \Gamma. \quad (11.8)$$

By definition Γ is the sum of all partial widths over the open channels. Thus, the branching ratio for the decay into one definite channel $bB \equiv c'$ is equal to $\Gamma_{c'}/\Gamma$ and the Breit–Wigner cross section for the formation of the resonance via channel c and the decay via channel c' is

$$\sigma(E) = \frac{\pi}{k_{\mathrm{in}}^2} \cdot \frac{\Gamma_c \Gamma_{c'}}{(E - E_{\mathrm{R}})^2 + \Gamma^2/4}. \quad (11.9)$$

This derivation is simplified and must be carried out—when there is interference with a direct background contribution and for the description of a differential cross section via a partial-wave expansion near a resonance—with complex *scattering amplitudes*. For elastic s-wave scattering this results in a resonant scattering amplitude of the form:

$$A_{\mathrm{res}} = \frac{\mathrm{i}\Gamma_{aA}}{(E - E_{\mathrm{R}}) + \mathrm{i}\Gamma/2}. \quad (11.10)$$

When a direct background is present, in addition to the pure resonance term and the pure direct (smooth) term, a typical interference term appears, which may be constructive or destructive. For σ we have then:

$$\sigma_{\mathrm{tot}} = \left| A_{\mathrm{res}} + A_{\mathrm{pot}} \right|^2 = \sigma_{\mathrm{res}} + \sigma_{\mathrm{pot}} + 2\,\mathrm{Re}\left(A_{\mathrm{res}} A_{\mathrm{pot}}^* \right) \quad (11.11)$$

where A_{pot} is the amplitude of the weakly energy-variable potential scattering.

11.3 Derivation of the partial-width amplitude for nuclei (s-waves only)

The connection between the resonant scattering wave function and the wave function of the eigenstate of the nucleus is made using the *R-matrix theory*. Their basic features (for more details see [20]) are approximately:

- The two wave functions and their first derivatives are matched continuously at the nuclear radius (the edge of the potential or similar).
- The condition for a resonance is equivalent to the wave-function amplitude in the nuclear interior taking on a maximum value. This happens exactly if the matching at the nuclear radius occurs with a wave function with gradient zero (horizontal tangent).

Figure 11.2 illustrates the conditions for resonance. The two conditions may be summarized such that both logarithmic derivatives L (L_0 for pure s-waves) at the nuclear radius are exactly zero. With the form of the wave function in the external region

$$u_0(r) = \mathrm{e}^{-\mathrm{i}kr} - \eta_0 \mathrm{e}^{\mathrm{i}kr}, \quad r > a \tag{11.12}$$

and the wave numbers in the external k in the nuclear interior κ we obtain

$$L_0(E) = \left(\frac{a}{u_0} \frac{\mathrm{d}u_0}{\mathrm{d}r} \right)_{r=a}, \tag{11.13}$$

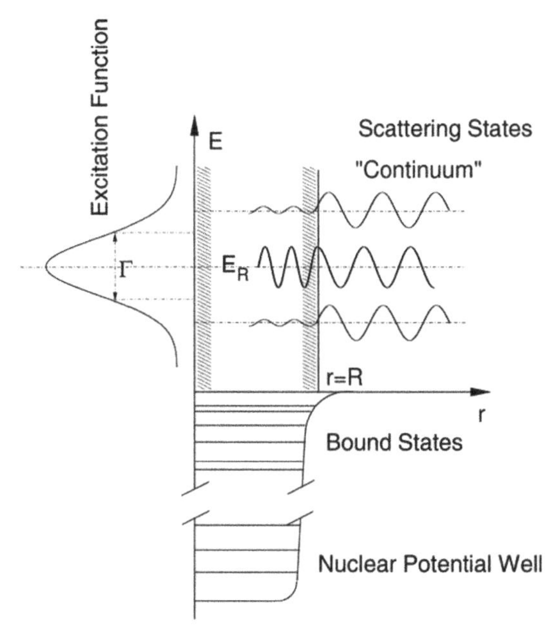

Figure 11.2. Boundary conditions at the nuclear (potential) surface for the appearance of a resonance in the excitation function.

which leads to the scattering function η_0 as a function of L_0:

$$\eta_0 = \frac{L_0 + \mathrm{i}ka}{L_0 - \mathrm{i}ka}\mathrm{e}^{-2\mathrm{i}ka}. \tag{11.14}$$

Inserting this scattering function into the known expressions for elastic scattering and absorption (and with $L_0 = \mathrm{Re}(L_0) + \mathrm{i}\ \mathrm{Im}(L_0)$) the result is

$$\sigma_{\mathrm{el}} = \frac{\pi}{k_{\mathrm{in}}^2}\left|1 - \eta_0\right|^2 = \frac{\pi}{k_{\mathrm{in}}^2}\left|\left[\mathrm{e}^{2\mathrm{i}ka} - 1\right] - \frac{2\mathrm{i}ka}{\mathrm{Re}\,L_0 + \mathrm{i}(\mathrm{Im}\,L_0 - ka)}\right|^2 \tag{11.15}$$

and

$$\sigma_{\mathrm{abs}} = \frac{\pi}{k_{\mathrm{in}}^2}\left(1 - \left|\eta_0\right|^2\right) = \frac{\pi}{k_{\mathrm{in}}^2}\left[\frac{-4k_{\mathrm{in}}a\,\mathrm{Im}\,L_0}{(\mathrm{Re}\,L_0)^2 + (\mathrm{Im}\,L_0 - ka)^2}\right]. \tag{11.16}$$

By expanding Re L_0 in a Taylor series and terminating it after the first term, by comparison—in addition to obtaining the resonance scattering amplitude (with $A_{\mathrm{pot}} \propto \mathrm{e}^{2\mathrm{i}ka} - 1$)—one obtains the results:

$$\sigma_{\mathrm{el}} = \frac{\pi}{k_{\mathrm{in}}^2}\left|\left(\mathrm{e}^{2\mathrm{i}ka} - 1\right) + \frac{\mathrm{i}\Gamma_{aA}}{\left(E - E_{\mathrm{R}}\right) + \mathrm{i}\Gamma/2}\right|^2 \tag{11.17}$$

$$\sigma_{\mathrm{abs}} = \frac{\pi}{k_{\mathrm{in}}^2}\frac{\Gamma_{aA}(\Gamma - \Gamma_{aA})}{(E - E_{\mathrm{R}})^2 + \Gamma^2/4}. \tag{11.18}$$

One sees that in agreement with our definition of absorption this encompasses all exit channels except the elastic channel.

11.4 The first evidence of resonant nuclear reactions

Around 1934/1935 two models of nuclear interaction were put forward and debated:

- The *single-particle model*, which describes the interaction of a nucleon with nuclei as a global interaction with each nucleus as a whole with very little regard for the internal structure. Elastic processes should therefore depend only weakly on the mass number of the target nuclei, predominantly via the radius dependence with A, and also on the incident energy, a concept realized in the optical model, see chapter 10. Inelastic processes should be fast, peripheral and involve only very few nucleons.
- The *compound-nucleus model* relies on the assumption that an incident nucleon (or other nucleus) is absorbed by the target nucleus, shares its properties such as mass, energy and spin with all nucleons, and may be emitted after a long delay by thermal excitation (evaporation, compound-elastic scattering) or another nucleon or a group of nucleons is evaporated in the same way (compound-nuclear (CN) reaction).

The applicability of each model could only be decided experimentally. E P Wigner gives a good account of these early developments. His statement, cited from [28],

'Experimental work constituted, in my opinion, the most important step in the development', expresses this.

11.5 Neutron resonances

In the years around 1934 Enrico Fermi and his collaborators (Amaldi *et al* [1–3, 13]) performed experiments in Rome with neutrons, especially slow neutrons that were produced by slowing down fast neutrons in hydrogenous materials. The fast neutrons were produced in (α, n) reactions on nuclei such as ^{9}Be with α particles from radium, radon, or polonium sources. Figure 11.3 shows a typical set-up of a neutron source surrounded by a paraffin moderator, a target absorber and a detector for capturing γ radiation. An energy variation was achieved by changing the temperature environment with which the neutrons became equilibrated. Their very systematic studies of capture reactions revealed very different cross sections for different elements. Very high absorption was observed for B and Cd, but only for very slow neutrons, also on many other nuclei without any systematics between neighboring nuclei. Other groups also studied the interaction of neutrons with nuclei. Other key experiments are those of Bjerge *et al* [5], Dunning *et al* [11], Szilard [24] and Moon *et al* [21, 22]. In these experiments the 'selective absorption' of slow neutrons by different nuclei was investigated by studying the emission of γ after absorption of the neutrons as a function of energy and for different nuclei. One has to keep in mind that neutron physics was difficult in many respects (e.g. in determining their energies, making sure that the neutrons were really thermal, their proper collimation, background shielding and their detection). Nevertheless the results were that absorption cross sections differed so

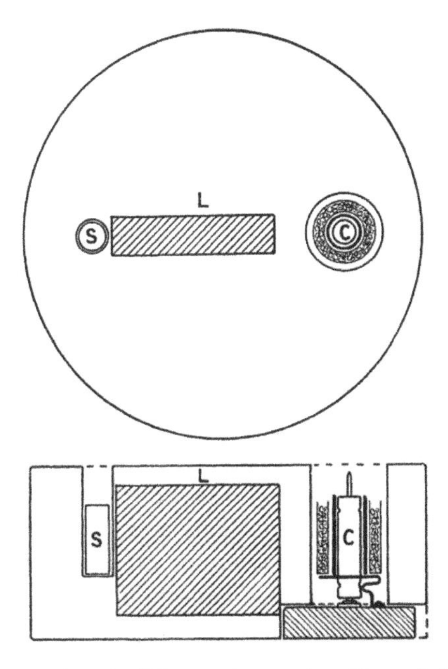

Figure 11.3. Typical set-up for experiments of slow neutron capture. Shown is the Po/Be source S in a paraffin block, a lead absorber shield for primary γ L and the detector C, here a Geiger–Müller counter surrounded by the absorber target material [2].

widely between different nuclei that a simple absorption law $\sigma_{\mathrm{abs}} \propto 1/v$ (as evidenced for fast neutrons as they were produced by the Rn–Be source) could not be valid. Cross sections of $\sigma_{\mathrm{abs}} \approx 3000$, 7000, and 10 000 barn, respectively, were reported for B, Y and Cd, about 1000 × larger than for fast neutrons. The idea that with slow (thermal or sub-thermal) neutrons an $A + 1$ nucleus formed in the neutron capture process became prevalent, but no explanation of the differing cross sections and their energy dependence emerged.

Leo Szilard, after another experiment on an In absorber [24], already concluded that 'it would therefore seem that some elements have fairly sharp regions of strong absorption...'. This description fits the properties of what were soon called *resonances* in the sense described above, but for neutrons they were still described as 'selective absorption' or 'selective capture'.

Experimental progress, particularly in the creation of high energy resolution neutron beams (see also figure 12.8 in chapter 12), was decisive in many respects:

- The exact Breit–Wigner form plus interference with direct background could be established.
- The very rich structure of low-lying CN energy levels could be studied.
- The statistics of level densities and level widths could be established.
- Near the onset of overlapping of levels, phenomena such as statistical Ericson fluctuations [12] and finally, at still higher excitations, the smooth CN behavior described by Hauser and Feshbach [16] in the *statistical model* appeared, which had to be distinguished from smooth direct-interaction amplitudes, see e.g. [26, 27].

An example of the resonance behavior that also has practical importance in the slowing down of reactor neutrons in ^{238}U is shown in figure 11.4.

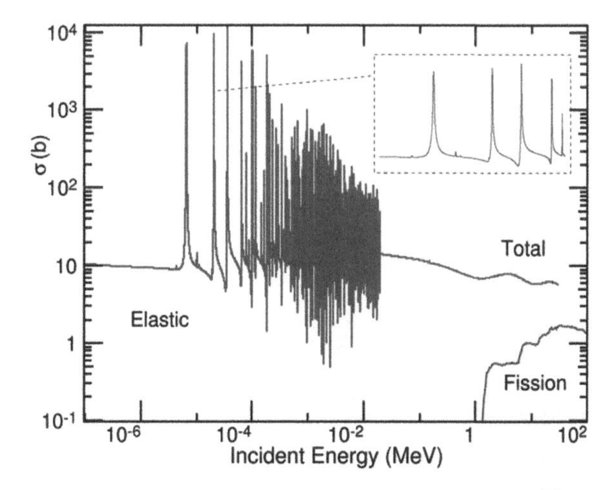

Figure 11.4. Excitation function of the cross section of neutrons interacting with ^{238}U. At low energies elastic scattering dominates the total cross section with interference with narrow compound-nucleus resonances. At higher energies the region of resonances overlaps, then a smooth cross section is visible and other channels open up, such as fission. Adapted from [23].

11.6 Charged-particle resonances

Around the same time as neutron absorption by nuclei, in particular using thermal neutrons, was performed, similar studies, with protons in particular, became feasible with the development of accelerators with higher voltages than the first (Cockroft–Walton) machine. In order to surmount the Coulomb barrier before being absorbed by nuclei the protons had to have energies in the MeV range. Pioneering work was performed at the Department of Terrestrial Magnetism (DTM) of the Carnegie Institution in Washington, DC, by Tuve *et al* and at Madison by Herb *et al.* Due to the high energy resolution and stability of the beams from these new accelerators, they were able to measure excitation functions that showed the resonances nicely [14, 15, 25]. As an example, the 2 m open-air Van de Graaff accelerator at DTM that reached 1.3 MeV proton energy is shown in figure 11.5, together with an excitation function of proton scattering from carbon as function of energy. In 1935 R Herb [17] developed the first Van de Graaff accelerator in a pressurized vessel thus providing much higher breakthrough voltages from the accelerator terminal to ground and thus higher maximum voltages, here 2 MV [18], as shown in figure 11.6. This design was the prototype for the many later Van de Graaff accelerators. A large number of nuclei was investigated with protons via the capture reaction. The de-excitation γ rays were registered with an electroscope as shown in figure 11.7. Figure 11.8 shows excitation

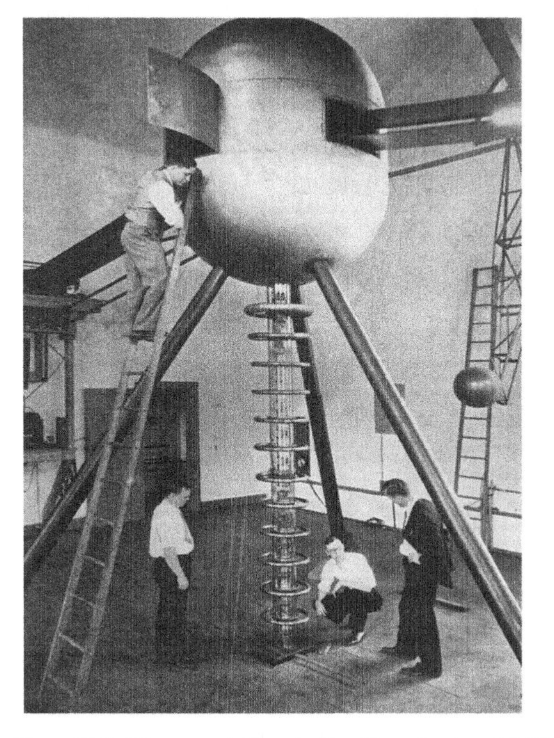

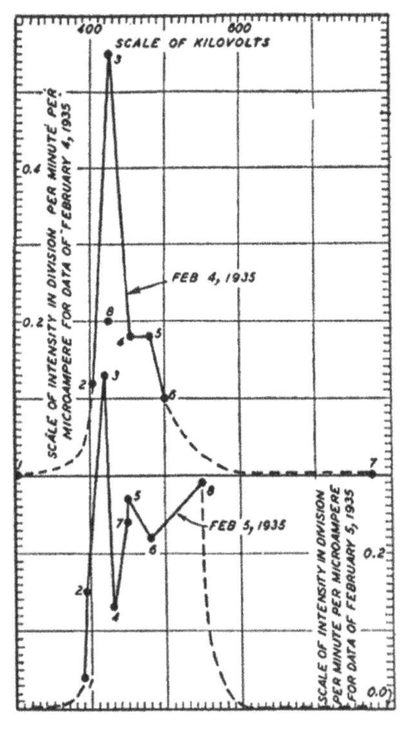

Figure 11.5. The open-air Van de Graaff accelerator at DTM (with paper charging belt) and an excitation function of protons on C showing early evidence of CN resonances in the compound system [14, 25]. Copyright 1935 American Physical Society.

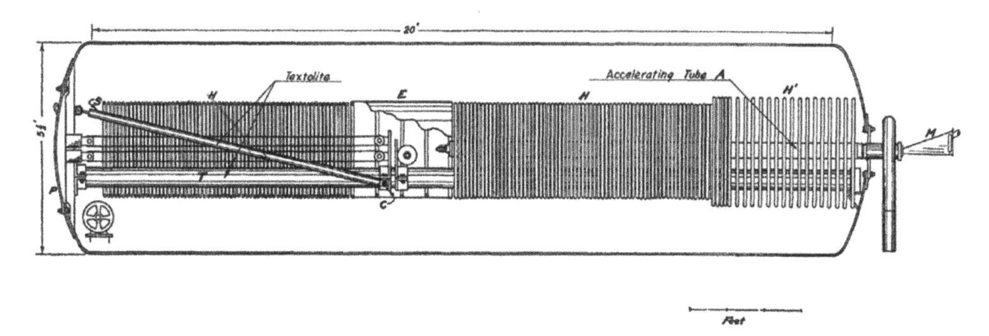

Figure 11.6. The 1937 Madison Van de Graaff accelerator enclosed in a pressure tank. Reproduced with permission from [18]. Copyright 1937 American Physical Society.

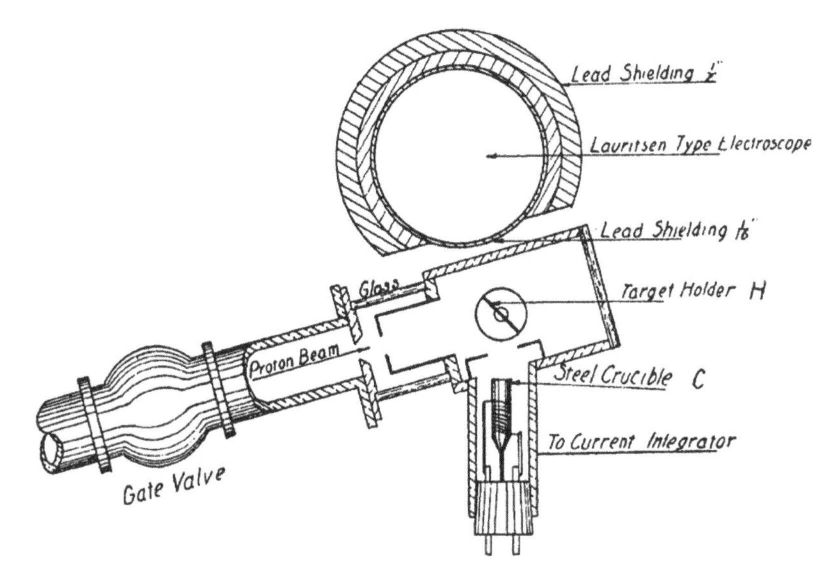

Figure 11.7. The Madison reaction chamber for (p,γ) capture experiments. Reproduced with permission from [19]. Copyright 1937 American Physical Society.

functions of protons on Li and F targets that clearly exhibit resonance structures, which could be identified with CN states in the corresponding nuclei. Their observation required a sufficiently good energy resolution for the accelerator as well as sufficiently thin targets [19].

11.7 The compound-nucleus model

Niels Bohr [6–9] used the observations of 'selective effects, selective capture, selective absorption' of low-energy neutrons on the one hand and 'charged-particle resonances' on the other to formulate his model of the formation of a *compound nucleus* explaining the properties of the high cross sections as resonances in the system of the target nuclei plus one neutron being captured to form a highly excited, rather long-lived nuclear state (lifetimes up to 10^6 times longer than the traversal time). Among the properties of the compound nuclei was their decay independent of their

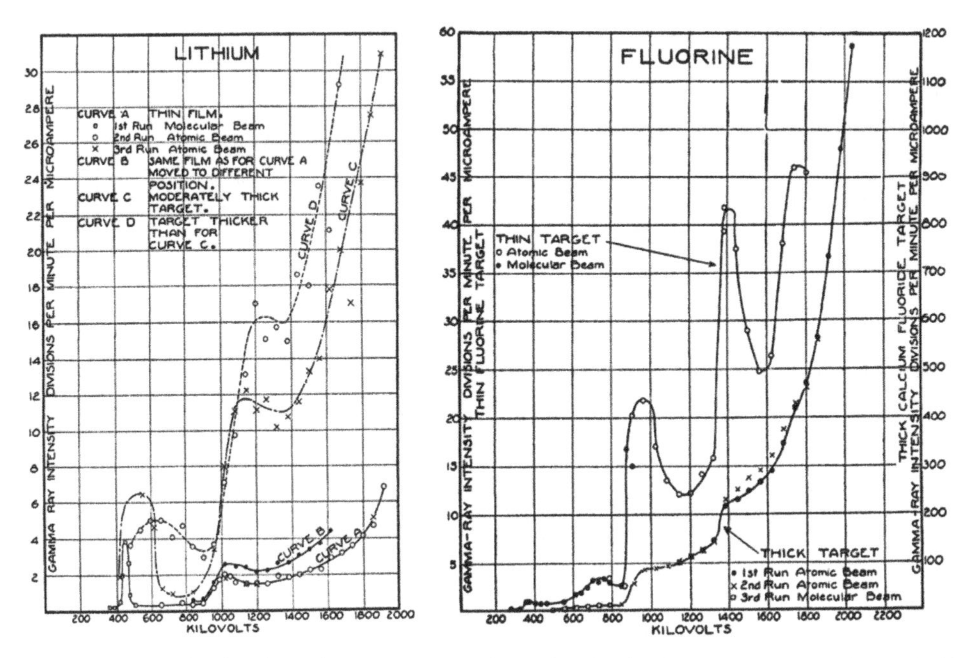

Figure 11.8. Excitation functions of protons scattered from Li and F showing CN resonances. Adapted with permission from [19]. Copyright 1937 American Physical Society.

Figure 11.9. Bohr's illustration of the compound-nucleus process. Reproduced with permission from [7]. Copyright 1936 Nature Publishing Group.

formation (*Bohr's independence hypothesis*). Bohr presented the famous drawing explaining the CN interaction between a nucleon and a target nucleus [7], depicted in figure 11.9. Bohr [6–9], Breit and Wigner [10], and Bethe [4] were the main early proponents of the compound-nucleus model of nuclear reactions.

Bibliography

[1] Amaldi E and Fermi E 1936 *Ric. Sci.* A **6** 544
[2] Amaldi E, D'Agostino O, Fermi E, Pontecorvo B, Rasetti F and Segrè E 1935 *Proc. R. Soc.* A **149** 522
[3] Amaldi E and Fermi E 1936 *Ric. Sci.* **1** 310
[4] Bethe H A 1937 *Rev. Mod. Phys.* **9** 69
[5] Bjerge T and Westcott C H 1935 *Proc. R. Soc.* A **150** 709
[6] Bohr N 1936 *Nature* **137** 344
[7] Bohr N 1936 *Nature* **137** 351
[8] Bohr N and Kalckar F 1937 *Mat.-Fys. Medd. K. Dan. Vidensk. Selsk.* **27** no. 10
[9] Bohr N 1937 *Science* **86** 161
[10] Breit G and Wigner E 1937 *Phys. Rev.* **49** 519
[11] Dunnning J R, Pegram G B, Fink G A and Mitchell D P 1935 *Phys. Rev.* **48** 265
[12] Ericson T and Mayer-Kuckuk T 1966 *Ann. Rev. Nucl. Sci.* **16** 183
[13] Fermi E, Amaldi E, D'Agostino O, Rasetti F and Segrè E 1934 *Proc. R. Soc.* A **146** 483
[14] Hafstad L R and Tuve M A 1935 *Phys. Rev.* **48** 306
[15] Hafstad L R, Heydenburg N P and Tuve M A 1936 *Phys. Rev.* **50** 504
[16] Hauser W and Feshbach H 1952 *Phys. Rev.* **87** 366
[17] Herb R G, Parkinson D B and Kerst D W 1935 *Rev. Sci. Instrum.* **6** 261
[18] Herb R G, Parkinson D B and Kerst D W 1937 *Phys. Rev.* **51** 75
[19] Herb R G, Parkinson D B and Kerst D W 1937 *Phys. Rev.* **51** 691
[20] Lane A M and Thomas R G 1958 *Rev. Mod. Phys.* **30** 145
[21] Moon P B and Tillman J R 1935 *Nature* **135** 904
[22] Moon P B and Tillman J R 1936 *Proc. R. Soc.* A **153** 476
[23] National Nuclear Data Center 2012 Brookhaven National Laboratory
[24] Szilard L 1935 *Nature* **136** 950
[25] Tuve M A, Hafstad L R and Dahl O 1935 *Phys. Rev.* **48** 315
[26] Vogt E 1968 *The Statistical Theory of Nuclear Reactions* (*Advances in Nuclear Physics* vol 1) (New York: Plenum) p 261
[27] Vogt E 1972 *Rev. Mod. Phys.* **34** 723
[28] Wigner E 1955 *Am. J. Phys.* **23** 371

Chapter 12

Nuclear reactions and tests of conservation laws

Depending on the strength of the interaction a number of conservation laws and their violation have been formulated. Table 12.1 lists all possibilities and relates them to various operators of the nucleon–nucleon interaction. It is evident that the number of violations increases with increasing weakness of the interaction. It is also evident that each conservation law has to be investigated separately for each interaction. For the weak interaction the complete violation of the parity symmetry corresponding to the invariance of physical processes against the mirror operation about the origin was the most conspicuous and unexpected phenomenon (the 'Wu

Table 12.1. Conservation quantities and their violation, and fundamental interactions. Conservation: +; violation: −.

Conservation quantity or symmetry	Strong	Electromagn. interaction	Weak
Mass *m*/energy *E*			
Momentum *p*	+	+	+
Angular momentum *L*, *S*			
Charge *Q*	+	+	+
Isospin *T*	+	−	−
Strangeness s			
Charm c	+	+	−
Beauty b, topness t			
Parity *P*	+	+	−
Charge conjugation *C*	+	+	−
Baryon number *B*	+	+	+
Lepton number *s*		+	+
Hypercharge *Y*	+	+	−
Time reversal *T*	+	+	−
Charge parity *CP*	+	+	−
CPT	+	+	+

experiment') and triggered a host of further investigations of all possible conservation laws. In addition to the static properties of particles, nuclei and atoms, nuclear reactions are a tool to study the effects of possible violations.

12.1 The first tests of parity violation in hadronic reactions

Which observables are sensitive to parity violation? The parity operation entails

$$\begin{aligned}
\mathbf{P}\vec{r} &= -\vec{r} \\
\mathbf{P}r &= -r \\
\mathbf{P}\theta &= \pi - \theta \\
\mathbf{P}\phi &= \pi + \phi \\
\mathbf{P}t &= t \\
\mathbf{P}\vec{p} &= -\vec{p} \\
\mathbf{P}\hat{L} &= \hat{L} \\
\mathbf{P}\hat{S} &= \hat{S}.
\end{aligned}$$

Under *parity conservation* the physics should not change under the mirror operation, conversely under *parity violation*. Thus, any physical quantity containing odd orders of $\vec{r}$ such as the electric dipole moment of a particle $\vec{\mu}_e = \langle q \cdot \vec{r} \rangle$ is sensitive to parity violation.

The earliest attempt at measuring parity violation in a nuclear reaction was performed by Tanner [51] by comparing the ^{19}F(p,α_0)^{16}O reaction at $\theta = 0°$ with ^{19}F(p,$\alpha_1\gamma$)^{16}O yield across a resonance in ^{20}Ne at $E_p = 340$ keV. Under parity conservation the yield of the first reaction is predicted to be zero. In fact, it showed no sign of the resonance that was very clear in the second one, as shown in figure 12.1. Thus, to an upper limit of $1 \cdot 10^{-4}$ no indication of a parity-violating amplitude was detected. Note that the weak contribution to the hadronic interaction is expected to be on the 10^{-7} level, except when some enhancement mechanism

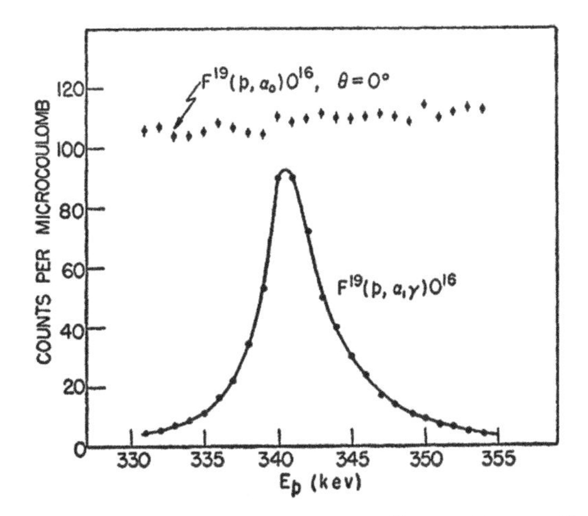

Figure 12.1. Plot of the excitation functions of the reactions ^{19}F(p,α_0)^{16}O and ^{19}F(p,$\alpha_1\gamma$)^{16}O* across the resonance at $E_p = 340$ keV in ^{20}Ne. Reproduced with permission from [51]. Copyright 1957 American Physical Society.

amplifies this ratio (e.g. in interfering neutron resonances or when the *P*-invariant amplitude is small as in cross-section minima).

Pseudoscalar observables (expectation values of products of a polar and an axial vector) are also sensitive, such as the correlation of the momentum vector of the electrons emitted from a ^{60}Co source and the spin of the Co nuclei aligned by a strong magnetic field in the famous Wu experiment.

In nuclear reactions the appearance of a longitudinal spin polarization, i.e. a polarization of the exit-channel particles along the direction of emission (usually designated as the z' axis) is only possible under parity violation. *Polarization* is defined as the expectation value of a spin operator. Similarly the corresponding longitudinal *analyzing power* A_z, the response of a nuclear reaction to the longitudinal polarization of an incident beam, has to disappear under parity conservation.

At Los Alamos, using the 15 MeV proton beam of an FN tandem Van de Graaff accelerator with the LANL Lambshift polarized ion source [38] the first such experiment was undertaken to look for parity violations in proton–proton elastic scattering (later also in $\vec{\mathrm{p}}$-^{4}He elastic scattering). Assuming that in the strong and Coulomb interactions parity is conserved, only the weak-interaction contribution could be responsible for a violation and very small effects, i.e. values of A_z, could be expected. This turned out to be the case. Normally the transverse polarization of the incident beam introduces an azimuthal dependence into the cross section that is normally φ independent. In the case of a longitudinal polarization there cannot be a φ dependence. Therefore, a detector design that integrates over all azimuthal events could be used. The total cross sections of events with the polarizations of the incident protons alternating in the forward and the backward directions served as a measure of the longitudinal analyzing power. The two images of figure 12.2 show the (quite simple) scheme of the experiment and details of the detector apparatus. In order to obtain the necessary statistics single scattered protons were not counted, instead their current produced in a 4π scintillator–photomultiplier arrangement was integrated and normalized to the incident proton beam current. The final result was $A_z = (-1.7 \pm 0.8)$, a clear signal of parity violation.

Similar, but more refined, experiments of scattering of longitudinally polarized protons from unpolarized protons followed at different laboratories, listed in table 12.2. Figure 12.3 shows the data together with the Desplanques, Donoghue and Holstein (DDH) prediction. The smallness of the effect required not only good statistics, i.e. long running times under stable experimental conditions, but also careful evaluation of systematic errors (such as from small transverse polarization components). The Desplanques, Donoghue and Holstein (DDH) model [17] is based on describing the NN interaction by meson exchange where one vertex is strong, the other weak and parity-violating. The weak interaction has been parametrized by six weak meson–nucleon coupling constants describing π, ρ and ω exchanges. Different experiments, among them the one described here, would in principle allow the determination of each of them separately. So far the two constants h_ρ^{pp} and h_ω^{pp} have been constrained by the pp experiments. For details see the latest review of [28] and references therein. In [32], in addition to a history of hadronic parity violation, new approaches in the framework of model-independent effective-field theories replacing the ‘classical’ DDH comparisons are discussed.

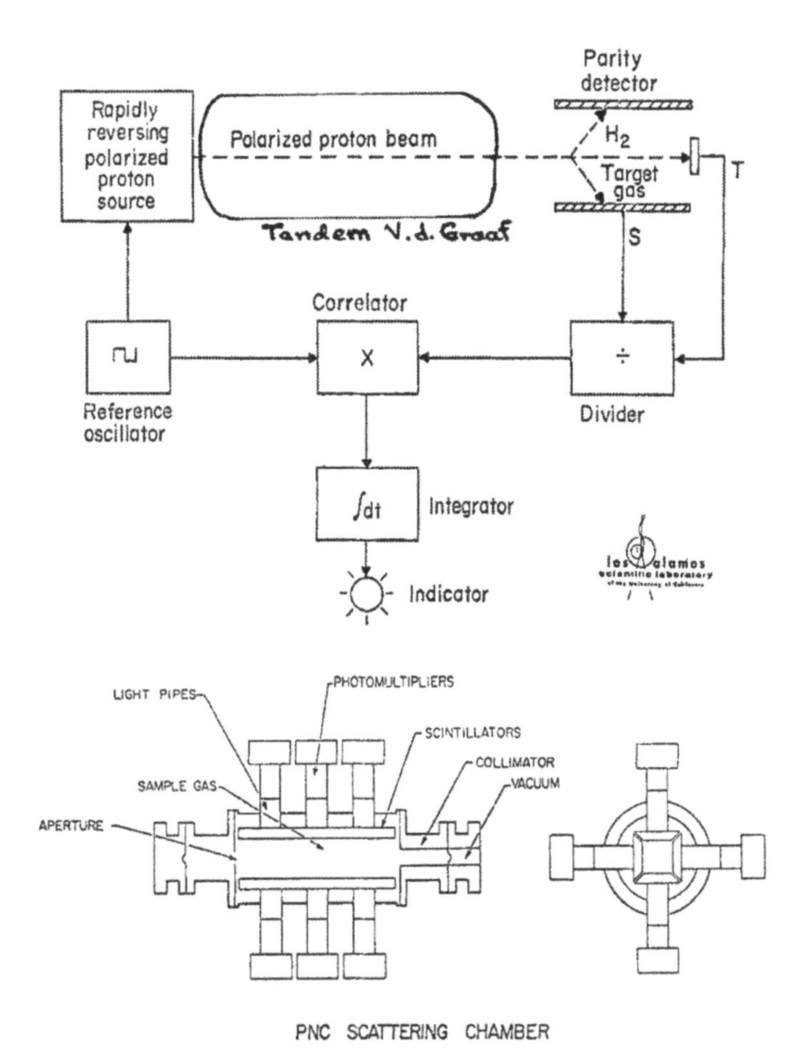

Figure 12.2. Set-up of the first hadronic parity-violation experiment at Los Alamos. Details of the 4π detector are shown in the lower half. Reproduced with permission from [42]. Copyright 1979 American Institute of Physics Publishing LLC.

Table 12.2. Results for the longitudinal analyzing power A_z of the $\vec{p}$p scattering experiments.

$E_{\mathrm{p,lab}}$ (MeV)	$A_z \times 10^7$	Laboratory	Year	Reference
13.6	$-(1.5 \pm 0.5)$	Bonn	1990	[19]
15.0	$-(1.7 \pm 0.8)$	LANL	1978	[42]
45	$-(3.2 \pm 1.1)$	SIN/PSI	1980	[6]
45	$-(2.21 \pm 0.84)$	SIN/PSI	1984	[7]
45	$-(1.5 \pm 0.22)$	SIN/PSI	1987	[34]
221	$+(0.84 \pm 0.29 \pm 1.7)$	TRIUMF	2001	[9]
800	$+(2.4 \pm 1.1)$	LANL	2001	[55]
5300	$+(26.5 \pm 6 \pm 3.6)$	ANL	1986	[37]

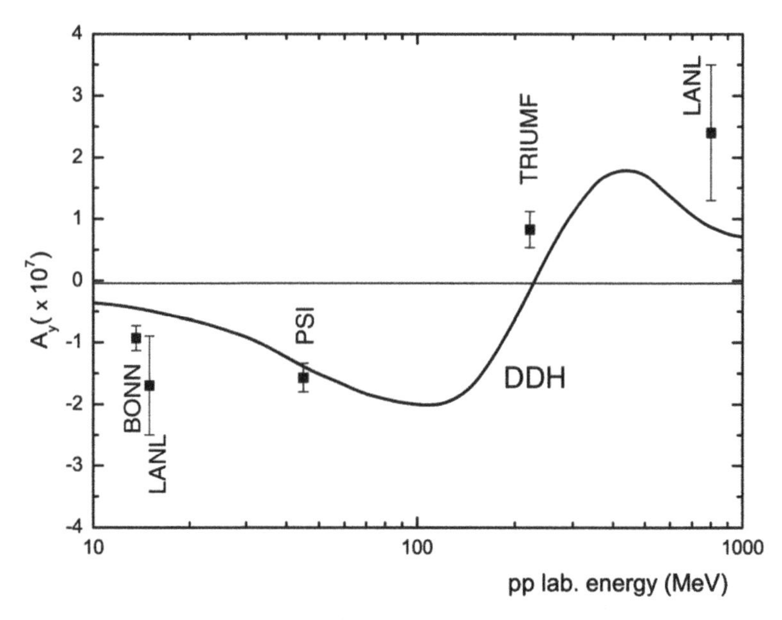

Figure 12.3. Measured values (see table 12.2) of the longitudinal analyzing power in pp scattering at 13.6 MeV (Bonn), 15 MeV (Los Alamos), 45 MeV (SIN/PSI), 221 MeV (TRIUMF) and 800 MeV (LANL), in comparison to predictions of the standard model (DDH).

12.2 First time-reversal tests

The CPT theorem (stating the invariance of physical processes and systems under the combined operations of charge conjugation *C*, the parity operation *P* and time reversal *T*) is one of the most fundamentally accepted theorems in physics. Its validity must and can, however, be investigated directly as well as by checking the three operations, *C*, *P* and *T*, separately. After a complete violation of *P* and *C* symmetries had been shown for the weak interaction, the validity of invariance under the combined *CP* operation was assumed, but only until a weak violation of *CP* was discovered for $K^0 - \bar{K}^0$ and later also for other systems. This immediately aroused interest in independently checking the time-reversal invariance *T*.

Time reversal is somewhat special because—unlike the other symmetries—there is no conserved quantity (quantum number) connected with it (for more detailed discussions see [21, 44]). The reason is the special nature of the operator of time reversal: it is *anti-linear* and *unitary* = *anti-unitary* and acts on operators as

$$\begin{aligned}\mathbf{T}t &= -t\\ \mathbf{T}\vec{r} &= \vec{r}\\ \mathbf{T}\vec{p} &= -\vec{p}\\ \mathbf{T}\hat{L} &= -\hat{L}\\ \mathbf{T}\hat{S} &= -\hat{S},\end{aligned}$$

which forbids certain static observables such as the electric dipole moment. In nuclear reactions time reversal is usually interpreted as reversal of motion, relating

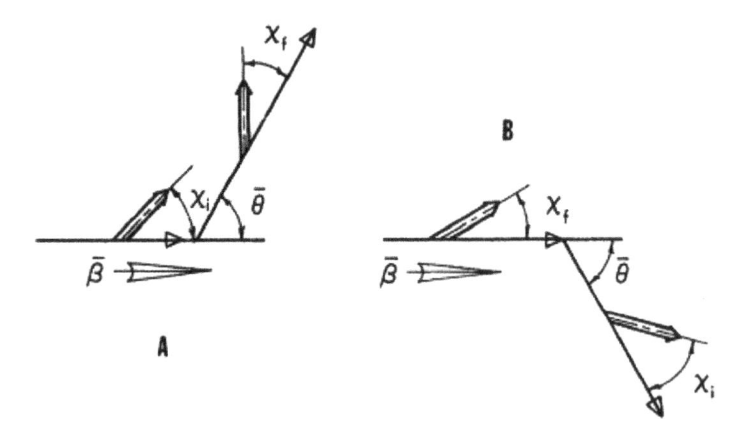

Figure 12.4. Situation of incident and exit momenta and polarization vectors before and after the time-reversal operation of reversing spins and momenta in A leading to situation B in the c.m. system. The angles χ_i and χ_f are arbitrary. The situations must be Lorentz-transformed to describe them in the laboratory system. Reproduced with permission from [25]. Copyright 1967 American Physical Society.

the input channel (aA) of a reaction with the exit channel (bB). For elastic scattering both are identical. There are three main possibilities to test the time-reversal invariance in nuclear reactions:

- By performing the time-reversal operation directly on momentum and spin vectors of a nuclear reaction, then—after an additional rotation—the physics of the reaction and its inverse should be the same. This is illustrated for the center-of-mass (c.m.) system of proton–proton scattering in figure 12.4. The quantities measured were the pp triple-scattering (*Wolfenstein parameters*) A and R for the two situations A and B in figure 12.4, using wire spark chambers to register the protons. Under time-reversal invariance the relation

$$\tan\theta = \frac{A + R'}{A' - R} \tag{12.1}$$

holds where θ is the laboratory scattering angle, A and A' are the transverse and longitudinal final polarizations, respectively, for an 100% longitudinally polarized incident beam, and R and R' are the same parameters for a 100% transversely polarized initial beam [48]. For an initial polarization of unity in situations A and B we obtain final polarizations P_{A} and P_{B}

$$\begin{aligned}
P_{\mathrm{A}} &= \left(R\sin\chi_i + A\cos\chi_i\right)\sin\left(\chi_f + \theta\right) \\
&\quad + \left(R'\sin\chi_i + A'\cos\chi_i\right)\cos\left(\chi_f + \theta\right), \\
P_{\mathrm{B}} &= \left(-R\sin\chi_f + A\cos\chi_f\right)\sin\left(\theta - \chi_i\right) \\
&\quad + \left(-R\sin\chi_f + A'\cos\chi_f\right)\cos\left(\theta - \chi_i\right).
\end{aligned}$$

The difference between P_A and P_B

$$P_A - P_B = \left[\left(A + R'\right)\cos\theta - \left(A' - R\right)\sin\theta\right]\sin\left(\chi_i + \chi_f\right)$$

vanishes if 12.1 holds. This comparison thus tests 12.1.

The experiment was performed at 430 MeV at the cyclotron of the University of Chicago. The difference between the two polarizations that should be equal if T is valid was

$$\Delta P = 0.0006 \pm 0.0028.$$

Thus, no violation of time-reversal invariance was found and only upper limits for its violation could be established.

- The (vector) analyzing power of a forward reaction is identical with the (vector) polarization of the inverse reaction, produced with unpolarized incident projectiles

$$\left(A_y\right)_{\rightarrow} = \left(P_y\right)_{\leftarrow} \tag{12.2}$$

with the caveat that the target spin must not be 0.

The earliest (key) experiments on possible time-reversal violations were the more difficult polarization/analyzing-power difference measurements. Oxley *et al* [43] had measured $P - A = 0.01 \pm 0.06$ in pp scattering but without explicit reference to time reversal. The first such experiments were undertaken in two labs simultaneously by Hillman *et al* in 1957/58 [31] and by Abashian *et al* [1]. One has to keep in mind that at the time no polarized ion sources were available and the polarization of proton beams and their measurement had to be provided by double or triple scattering, placing strong restrictions on the intensity and choice of angles and energies. Proton scattering on several nuclear targets was performed. The combined result of both experiments was

$$P - A = -0.014 \pm 0.014.$$

Later experiments also using beams from polarized ion sources did not reach the significance necessary to detect a time-reversal breaking.

- Violations of the principle of *detailed balance* which states that—if time-reversal invariance holds—up to phase-space factors (spin factors and momentum factors which cancel for elastic scattering) the cross section of a reaction is equal to the cross section of its inverse (taken at the same c.m. energy and scattering angle):

$$\frac{\left(\frac{d\sigma}{d\Omega}\right)_{\rightarrow}}{\left(\frac{d\sigma}{d\Omega}\right)_{\leftarrow}} = \frac{\left(2s_a + 1\right)\left(2s_A + 1\right)}{\left(2s_b + 1\right)\left(2s_B + 1\right)} \cdot \frac{k_{in}}{k_{out}}. \tag{12.3}$$

It is a question of fundamental importance which reaction mechanism should be chosen to obtain maximum sensitivity of the (weak) time-reversal violating

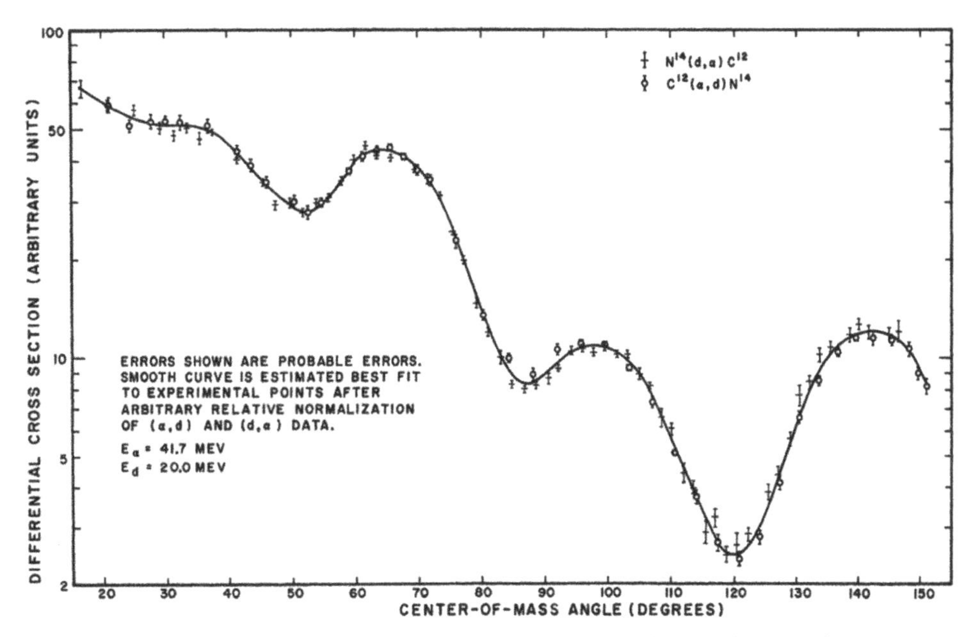

Figure 12.5. Plot of the angular distributions of the reactions $^{12}C(\alpha, d)^{14}N$ and $^{14}N(d, \alpha)^{12}C$. Reproduced with permission from [12]. Copyright 1959 American Physical Society.

reaction amplitude in the presence of a strong non-violating amplitude. The conclusion was that a complicated reaction with many channels such as compound processes would be better than simpler direct processes [29, 41]. A key experiment of this kind was the measurement by Bodanski *et al* in 1959 [12], see also [13, 53]. They chose the reaction

$$^{12}C(\alpha, d)^{14}N - 13.574 \text{ MeV}$$

with $E_{\alpha,\text{lab}} = 41.7$ MeV from the Washington cyclotron and its inverse. The results of both cross sections are shown in figure 12.5. It is obvious that the two sets of data agree qualitatively well. It is not quite easy to quantify the degree of (non-)agreement and the authors only estimated the non-invariant contribution to be <3%. Later the same group investigated the reactions $^{24}Mg + d \rightleftharpoons {}^{25}Mg + p$ with $<0.4\%$ non-invariance [13, 53].

Appreciably higher accuracy was obtained in experiments on the $^{24}Mg + \alpha \rightleftharpoons {}^{27}Al + p$ reactions [11, 52]. In this experiment in particular the deep minima of Ericson fluctuations were used for comparison. The authors gave an upper limit on time-reversal violation of $\approx 2 \cdot 10^{-3}$. For a discussion of the determination of the 'significance limit' of such experiments see [26, 27, 35].

- It is important for a clear signal of time-reversal violation to measure observables that are not simultaneously sensitive to parity violation, such as e.g. the electric dipole moment. Also, because the above tests require the comparison of two observables, a null-test experiment, i.e. the measurement of only one quantity, is desirable. One such observable has been identified by

Conzett [15, 16], the spin-dependent total cross section of the proton–deuteron system with the protons vector-polarized in the x-direction and the deuterons tensor-polarized along $y = z$, thus measuring the spin-correlation coefficient $C_{x,yz}$. Ideally such an experiment could be performed as a transmission-type experiment in a storage ring accelerator.

- For a long time the electric dipole moment of the neutron has been investigated with the idea that its electrical neutrality appears favorable in the interaction with electric fields. No time-reversal violation has been found so far, expressed by an upper limit of the dipole moment of $\mu_e \leqslant 2.9 \cdot 10^{-26}$ e · cm [4, 5].
- It should be mentioned here that in the future storage-ring accelerator experiments are planned to measure the electric dipole moment of the proton and the deuteron. They would make use of spin-polarized beams and the very weak interaction of the dipole moment with electric fields, summing up via the millions of revolutions of the beam in the ring. One such project (another at BNL, 2011) has been proposed by the JEDI collaboration at COSY-Jülich [18] with a planned upper limit of $\mu_e = 1 \cdot 10^{-28}$ e · cm. At that level new physics might appear.

12.3 The NN interaction and isospin

12.3.1 Generalities

Shortly after the discovery of the neutron 1932 and knowledge of its properties such as mass and spin Werner Heisenberg developed the concept of isospin (or isobaric spin) as a new symmetry. The most fundamental question connected with this concept is whether isospin is a conserved quantity and under which interactions it might be (eventually weakly) violated. Like in many other instances the study of this quantity in heavier nuclei (or their reactions) may be obscured by nuclear structure. Therefore naturally, first the reactions between pure nucleons were investigated, at very low energies. The *scattering length* is a useful concept to study isospin and its possible violation in the nucleon–nucleon or at most in nucleon–few-nucleon interactions.

Only a brief survey of the isospin operator (**T** or $\hat{T}$) properties will be given here (for a more detailed introduction to the field, see e.g. the resources listed in [44]).

- Formally it behaves just as the spin operator **S**: commutation rules, isospin raising or lowering operators, expectation values of $\hat{T}^2$ and T_3 (index 3 instead of z, because **T** has no spatial representation), $2T + 1$ substates forming isospin multiplets, etc.
- $\sqrt{T(T+1)}$ is an invariant scalar under isospin conservation, i.e. invariant under rotations in isospin space, i.e. $[H_{\text{strong}}, \mathbf{T}] = 0$ is the definition of isospin invariance, also charge independence. The value T is determined by the number of substates. T_3 measures the electric charge of the members of the multiplet with $Q = T_3 + Y/2$ with Y the *hypercharge* that for nuclei is the mass number A. Thus, the Coulomb interaction breaks isospin naturally. Non-trivial is, however, the possible breaking within the strong interaction. A number of examples in nuclear physics in general exhibit the approximately

fulfilled features required from isospin conservation but in detail show small indications of some breaking as well (at the 1% level):

- Energies of members of isospin multiplets, corrected for the simple Coulomb force, differ slightly ('Nolen–Schiffer anomaly')
- Isospin-forbidden nuclear reaction channels have weaker, but not zero transition strengths, compared to allowed channels ($^{12}C(p,p)^{12}C$ over a $T = 1/2$ and a $T = 3/2$ (forbidden) resonance in ^{13}N, the latter being also much narrower).
- For the reaction $d + d \rightarrow {}^4He + \pi^0$ that was believed to be truly isospin-forbidden, only recently was a very small cross section measured [40, 49].
- There are more examples.

Henley and Miller [30] have classified isospin breaking into four classes depending on the form of the isospin breaking potential in the two-nucleon (NN) systems. It is important to distinguish between charge independence (breaking) and charge symmetry (breaking). The connection between charge independence breaking, charge symmetry breaking and isospin invariance is: the charge symmetry operation consists of a 180° rotation about the two-axis $P_{CS} = \exp(i\pi T_2)$ with $[H_{strong}, P_{CS}] = 0$ when charge symmetry holds, and nn and pp observables should be equal.

12.3.2 The scattering length

The scattering length a was introduced in 1947 by Fermi and Marshall [20]. It is related to the integrated cross section extrapolated to $E \rightarrow 0$ which can be described by just one parameter. An extension of the theory, the effective range theory, improves the description with just one additional parameter r_{eff}. Both are essentially S-wave quantities. It is obvious that with only one parameter it is impossible to describe the details of any nuclear potential, but only an integral quantity independent of the specific interaction. The definition of a reads

$$a = \lim_{k\rightarrow 0} \frac{\tan \delta_0(k)}{k}$$

where $\hbar k = p$ is the beam momentum and δ_0 the S-wave scattering phase shift describing the reaction. For S-waves the extrapolated cross section is

$$\lim_{k\rightarrow 0} \sigma = 4\pi a^2.$$

In order to take into account the energy variation of the cross sections near $E = 0$ an expansion of σ as a function of the scattering length a was defined with a second parameter r_{eff}, the *effective range* where $\sigma = 4\pi [a(k)]^2$ with

$$\frac{1}{a(k)} = \frac{1}{a} - k^2 r_{eff}.$$

The NN system is fundamental, being free of nuclear structure effects and nevertheless the basis of all efforts to describe more complex nuclei, starting with the three-nucleon system in which three-body forces already come into play. Much progress has been achieved in describing such and higher-A systems with the precise NN interaction as the input, via Faddeev or effective-field theory techniques.

The NN system has one bound state, the deuteron (an np, $T = 0$, $^3S_1 + {}^3D_1$ mixture), the np spin singlet, $T = 1$ state $^1S_0$, as well as all other pp and nn states are unbound. However, from the point of view of isospin conservation, especially from charge independence observables such as np and nn, scattering lengths should be equal. The early low-energy experiments in the NN systems have been instrumental in determining the salient features of the NN interaction especially in exploring the isospin properties.

The np scattering length

For a_{np} np total cross sections had to be determined at the lowest possible energies. This was not hampered by a Coulomb barrier, but on the other hand special methods to work with neutrons of well-defined energies of sufficient intensities had to be developed. One method is to produce neutrons with a broad energy spectrum and select them in a reaction by their time of flight, the other to produce them with charged particle reactions such as ^{9}Be($\vec{d}$,n)^{10}B in a cyclotron. From the large number of early experiments to elucidate the properties of the NN interaction a few key experiments will be presented here.

- In the early experiments the total cross section of np scattering was measured and yielded large values but no detailed clues as to the nature of the interaction. A more realistic picture of the np interaction emerged after a strong spin dependence of the hadronic interaction was observed. The interaction differed so strongly between the triplet and the singlet states that the first is slightly bound (the deuteron) and the other exists only as unbound scattering states. Thus, from the definition and geometrical interpretation of the scattering length, both should have different signs. Already in 1937 large cross-section differences had been noticed in neutron scattering from ortho- and para-hydrogen, see [14, 24], a hint of the strong spin dependence of the np force, expressed by a spin–spin term in the nuclear potential.

  In order to measure this spin dependence, and also to directly obtain the signs of a, the coherent n–H_2 scattering was investigated on ortho- ($S_{H2} = 1$; parallel proton spins) and para-hydrogen ($S_{H2} = 0$; anti-parallel spins) by Sutton *et al* [50]. The relation between total ortho and para np cross sections and the scattering lengths is

$$\sigma(H_2, \text{ortho}) = 4\pi\left\{\left(\frac{a_s + 3a_t}{2}\right)^2 + \frac{1}{2}(a_t - a_s)^2\right\} \tag{12.4}$$

$$= \sigma(H_2, \text{para}) + 2\pi(a_t - a_s)^2. \tag{12.5}$$

The experiment consisted of a neutron beam from the reaction ^{9}Be($\vec{d}$,n)^{10}B (with deuterons from the 42″ Los Alamos cyclotron), moderated by paraffin and energy-analyzed by time of flight as shown in figure 12.8. At 20 K H_2 is to

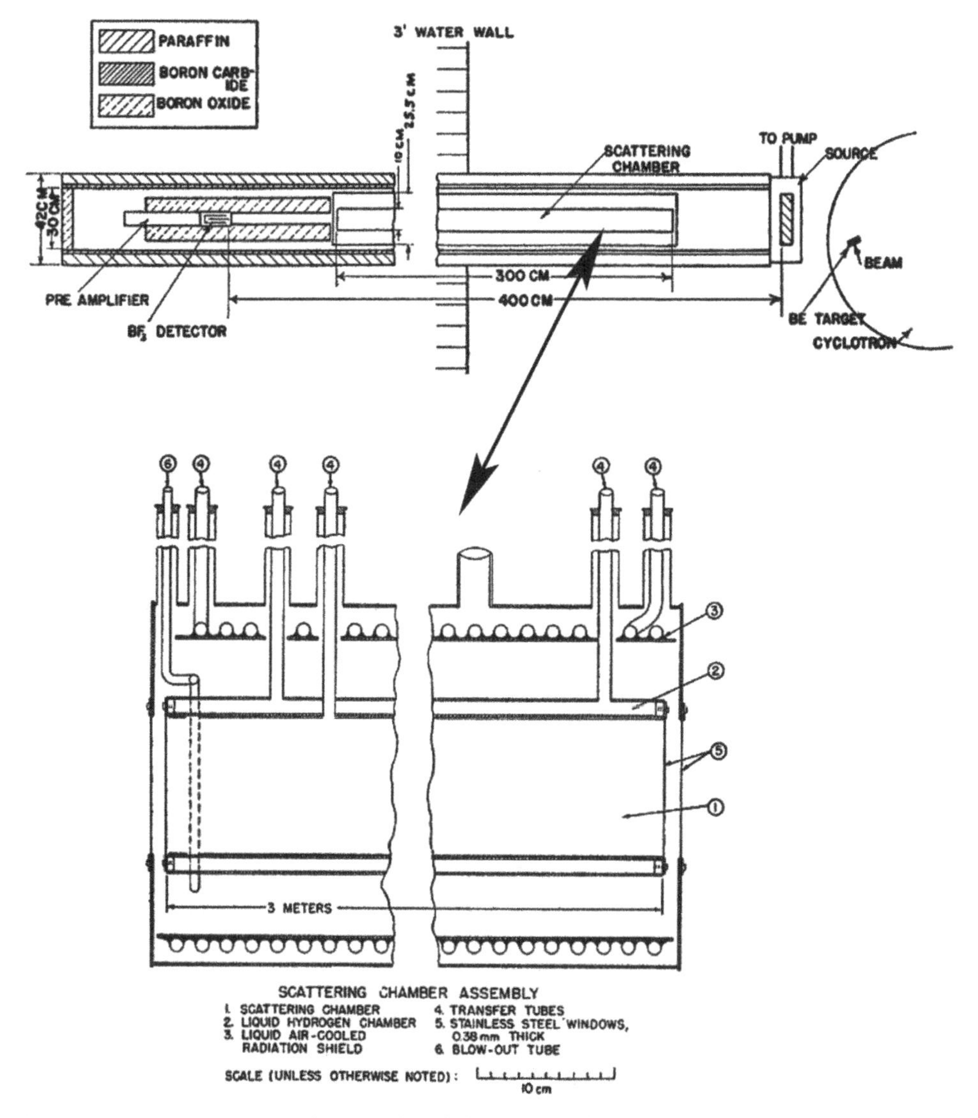

Figure 12.6. Experimental set-up for scattering of slow neutrons from H_2 with a liquid H_2 target vessel. The lower part is a detailed view of the scattering chamber. Adapted with permission from [50]. Copyright 1947 American Physical Society.

99.9% para-hydrogen (in thermal equilibrium after about 48 h), whereas pure ortho-H_2 cannot be prepared. However, H_2 is normally 75% ortho- and 25% para-H_2 and a cross-section difference must be evaluated. The essential special part of the set-up is thus a target vessel containing H_2, cooled by liquid hydrogen and sufficiently insulated, as shown in figure 12.6. A BF_3 counter was used as the neutron detector.

The neutron scattering cross sections on ortho- and para-H_2 are shown in figure 12.7 Careful evaluation of the data yielded the four observables shown

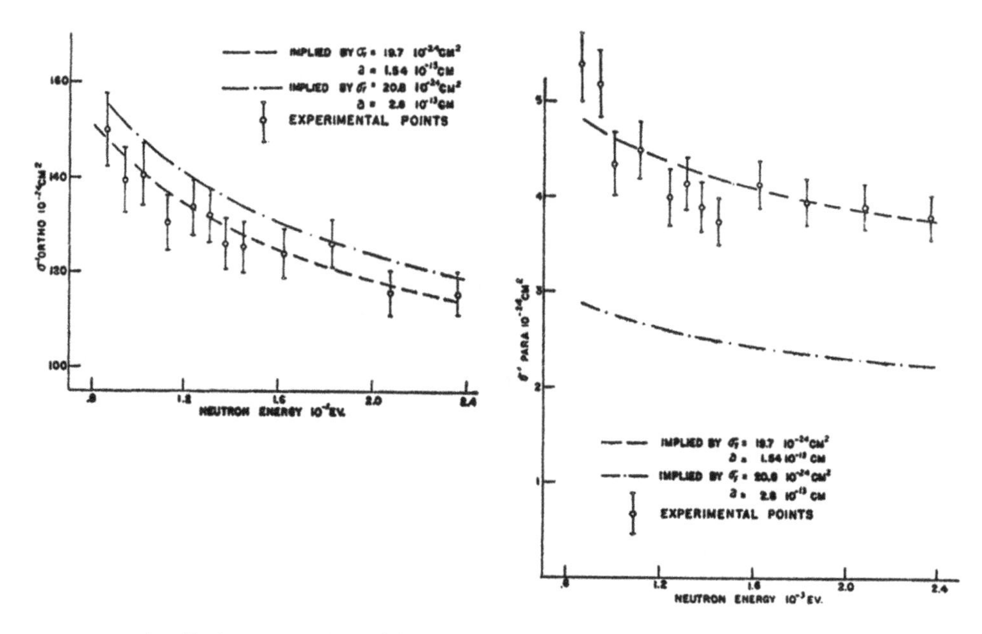

Figure 12.7. The effective cross sections of thermal neutron scattering on ortho- (left) and para-H_2 (right). The data clearly favor a total cross section, extrapolated to $E = 0$, of 19.7 b, together with a range of the nuclear force of 1.54 fm. Reproduced with permission from [50]. Copyright 1947 American Physical Society.

Table 12.3. np cross sections and scattering lengths from the early measurements of [50]. a_s and a_t are the singlet and triplet np scattering lengths, σ_s and σ_t the corresponding partial cross sections that have to be added incoherently with their statistical weights 1/4 and 3/4 to the total cross section σ_0.

	From [50]
a_s	−23.4 fm
σ_s	68.8 b
a_t	+5.2 fm
σ_t	3.6 b
σ_0	19.8 b

in table 12.3. The authors estimate the errors of their results to be about 5%, much improved over earlier results, thus making their measurements the first with reasonable errors allowing very important conclusions about the nature of the hadronic NN force to be made.

- A typical experiment allowing np measurements at neutron energies between 0.8–15 eV is that by Melkonian *et al* [39] using apparatus developed by Rainwater *et al* [45] and used for neutron cross-section measurements on a large number of different targets. Figure 12.8 shows the experimental set-up around the Columbia University cyclotron. Figure 12.9 shows the integrated

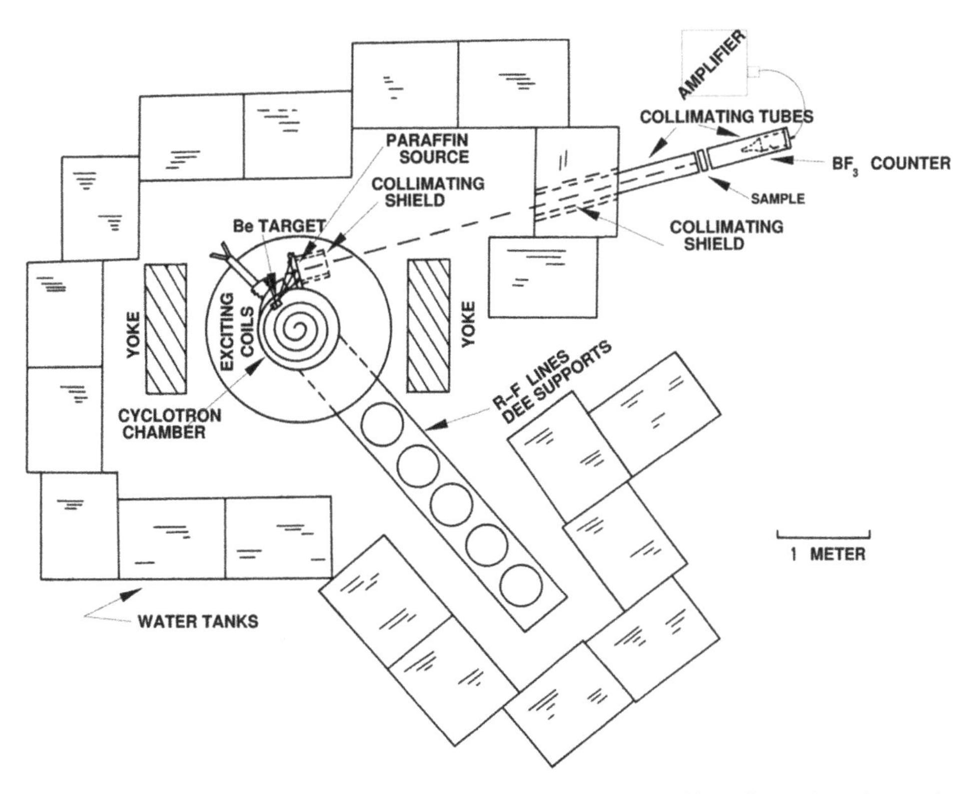

Figure 12.8. Set-up for the production of very slow neutrons at the Columbia cyclotron from the reaction ^{9}Be(d,n)^{10}B, after slowing down the neutrons in paraffin. The pulsed deuteron beam allows the selection of narrow energy intervals of the neutrons by μs gating and time-of-flight measurement along the 5.4 m flight path to the hydrogen target. The figure was redrawn from [45] for clarity. Copyright 1946 American Physical Society.

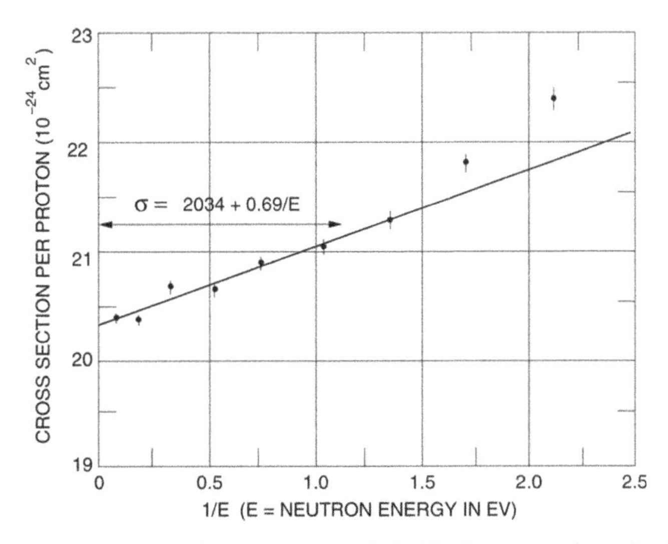

Figure 12.9. Result of the np cross-section measurement of [39]. The figure was redrawn for clarity. Copyright 1949 American Physical Society.

cross section as a function of $1/E$ fitted with the assumption that the cross section follows $\sigma = \sigma_0 + \beta/E_{\rm kin}$. σ_0 is the pure pn cross section extrapolated to zero energy. The extrapolation $E \to 0$ yielded

$$\sigma_0 = 20.36(10) \cdot 10^{-24}\,\mathrm{cm}^2$$

from which the absolute value of a scattering length of $a_{\rm np} = 12.7$ fm is deduced, but not the sign. The sign is either convention or must be obtained from interference terms in some observable. In any case, for a simple central repulsive potential, leading to unbound np states, a has a negative sign. The following conclusions were drawn:

- The np interaction is strongly spin dependent. It is repulsive in the triplet, attractive in the singlet S-wave state.
- The 'range' of the nuclear force is about 1.5 fm.
- The neutron is definitively a spin-1/2, not a spin-3/2 particle.

The pp scattering length

Charged-particle reactions require techniques different from neutron reactions, especially at low energies. The energy loss and finite range of protons require a scattering chamber with a good vacuum, and in the case of gas targets gas cells with very thin windows or windowless set-ups, such as filling the chamber with the target gas and replacing the entrance foil with an efficient differential pumping system (more modern versions have a target gas jet at much higher densities). Towards lower energies the Coulomb (Rutherford) cross section becomes very high and tends to obscure the nuclear part of the cross section which in turn is difficult to disentangle. pp scattering is no exception, and the identity of the entrance-channel particles leads to more complicated formulas (see also the next chapter 13), not only for the *Mott* cross section itself, but also in the interference terms with the pure nuclear part. However, we have to consider only S-wave nuclear scattering that is described by only a single phase shift δ_0. The relevant formula to be used when extracting the 'nuclear' pp scattering length is

$$\left(\frac{\mathrm{d}\sigma}{\mathrm{d}\Omega}\right)_{\rm Coul} = \left(\frac{Z^2e^2}{4E_\infty}\right)^2 \left\{ \frac{1}{\sin^4\frac{\theta}{2}} + \frac{1}{\cos^4\frac{\theta}{2}} - \frac{\cos\left(\eta_{\rm S}\ln\tan^2\frac{\theta}{2}\right)}{\sin^2\frac{\theta}{2}\cos^2\frac{\theta}{2}} - \frac{2}{\eta_{\rm S}}\sin\delta_0 \right.$$

$$\left. \times \left[\frac{\cos\left(\delta_0 + \eta_{\rm S}\ln\sin^2\frac{\theta}{2}\right)}{\cos^2\frac{\theta}{2}} + \frac{\cos\left(\delta_0 + \eta_{\rm S}\ln\cos^2\frac{\theta}{2}\right)}{\sin^2\frac{\theta}{2}} \right] + \frac{4}{\eta_{\rm S}^2}\sin^2\delta_0 \right\}.$$

pp scattering experiments were performed as soon as accelerators such as cyclotrons or Van de Graaff machines became available. Here two high-precision experiments from the 1950s will be described from which the pp scattering length was obtained [36, 54]. Earlier pp experiments have been surveyed in [10].

The nn scattering length

The problem with nn cross sections is the lack of a neutron target. One has to rely on targets such as deuterons or ^{3}He in which the effect of the neutron can be separated from the known reaction part without it. This means in a reaction such as n + d → p + n + n the kinematics of the exit channel can be chosen such that, e.g., the nn final-state interaction with a relative energy of E_{rel}(nn) → 0 can be studied. However, the binding of n and p in the deuteron causes the neutron to be only *quasi-free* requiring model-dependent assumptions (e.g. three-body-force effects) to extrapolate to the free neutron. Experiments of this type have been used to extract a_{nn}. Mostly kinematically incomplete measurements were performed in which only one outgoing particle was registered and the nn final-state interaction would show up as a peak in its spectrum. Kinematically complete experiments (two of the three outgoing particles are measured in coincidence) have the advantage that competing reactions and background can be clearly identified and separated. Early three-particle breakup reactions had to rely on relatively crude theoretical approaches (e.g. the Watson–Migdal theory) to obtain a best fit to the data and to extract low-energy parameters (for simple potential assumptions the depth and width and scattering lengths). Two early (key) experiments are those of Zeitnitz *et al* [56] using the D(n,p)2n reaction yielding $a_{nn} = -16.4 \pm 2.1$ fm and that of Baumgartner *et al* [8] using the ^{3}H(d,^{3}He)2n reaction yielding $a_{nn} = -16.1 \pm 1.0$ fm among a number of other reactions from different groups with results varying over a large range of values.

Modern theories are based on Faddeev or effective-field theory approaches and—especially for the final-state interaction—give excellent fits and thus quite reliable values for a_{nn}. Nevertheless the two most recent breakup experiments using somewhat different techniques [22, 33] yielded conflicting results of $a_{nn} = -16.1 \pm 0.4$ fm and -18.7 ± 0.6 fm, respectively. In view of the uncertainties about reactions with three interacting hadrons, another type of experiment where the nn interaction is a pure two-body interaction is believed to be more trustworthy. Two early examples are the D(π^-, γ)nn reaction [46, 47] which, with only the γ registered yielded a_{nn} between −15.1 and −19.1 fm, depending on the energy bin selection in the spectra, and the D (π^-, γ, nn) reaction [23] in which all three exit particles were measured and which yielded $a_{nn} = -16.4 \pm 1.3$ fm. Figure 12.10 shows the principle of the experiment. It should be mentioned that—in view of the inherent difficulties and unsatisfactory situation with the extraction of the nn scattering length—two exotic schemes for direct nn measurements have been proposed: one is to use the strong neutron flux in a pulsed high-flux reactor (YAGUAR), the other to use the neutron flux of a nuclear explosion, a one-shot experiment with destruction of the equipment.

Since these early key experiments the situation has not changed much. A summary of isospin breaking in the NN scattering lengths is shown in table 12.4 with the currently more or less accepted values. The early experiments as well as the new results clearly show isospin breaking, i.e. charge symmetry breaking and charge independence breaking.

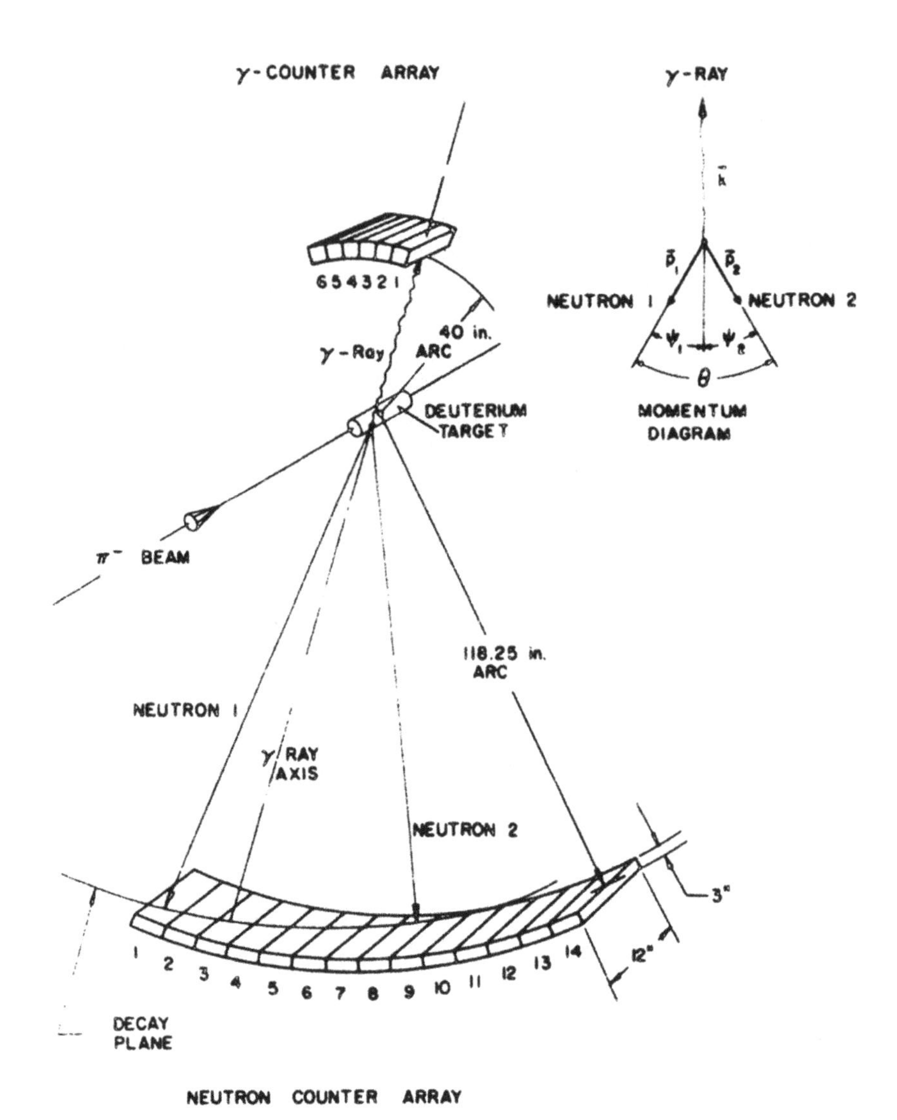

Figure 12.10. Experimental set-up and kinematics of the D(π^-, γ, nn) reaction. Reproduced with permission from [23]. Copyright 1965 American Physical Society.

Table 12.4. Currently accepted scattering lengths and effective ranges of the NN system.

NN	State	a_{NN} (fm)	r_{eff} (fm)
pp	$^1S_0$	-17.1 ± 0.2	2.85 ± 0.04
nn	$^1S_0$	-18.8 ± 0.3	2.75 ± 0.11
np	$^1S_0$	-23.715 ± 0.015	2.75 ± 0.05
np	$^3S_1$	$+5.423 \pm 0.005$	2.75 ± 0.05

12.3.3 Other reaction tests of isospin breaking

Other key experiments at higher energies and not based on scattering length were performed at the TRIUMF and IUCF laboratories in which the analyzing power from a polarized neutron beam on an unpolarized proton target was compared with the analyzing power of unpolarized neutrons on a polarized proton target. Under isospin invariance they should be the same but were found to be different: $\Delta A \equiv A_n - A_p = (34.8 \pm 6.2 \pm 4.1) \cdot 10^{-4}$, see [2, 3, 57], a clear manifestation of class IV charge symmetry breaking. The beauty of the experiments consists in the use of a zero crossing of the difference of analyzing power and polarization to enhance the sensitivity of the experiment. Figure 12.11 shows the scheme of the zero-crossing experiment and figure 12.12 the experimental set-up at TRIUMF.

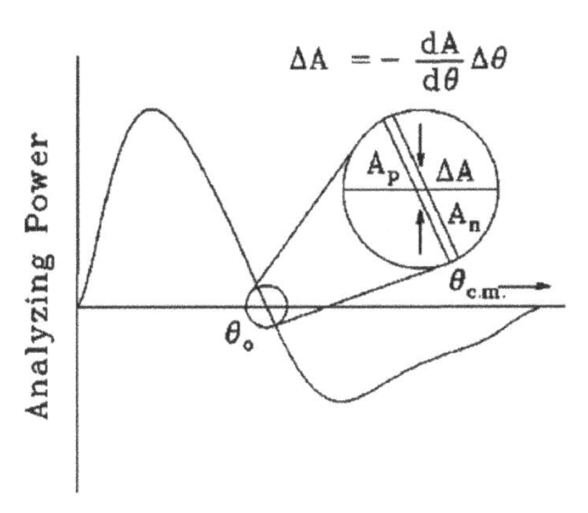

Figure 12.11. Scheme of the zero-crossing method of the analyzing power. Reproduced with permission from [57]. Copyright 1998 American Physical Society.

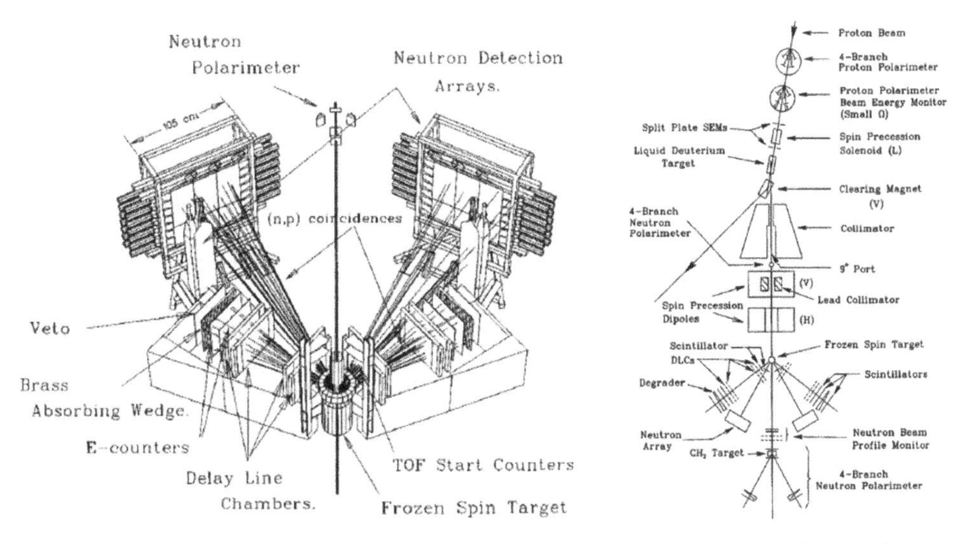

Figure 12.12. Detailed and general experimental set-up of the charge symmetry breaking experiment at TRIUMF. Reproduced with permission from [57]. Copyright 1998 American Physical Society.

With relatively good theoretical predictions it was possible to partly disentangle several different contributions to the charge symmetry breaking, among them predominantly the $\rho^0 - \omega$ mixing. With this improved knowledge of the microscopic sources of isospin breaking, an avenue for explaining the Nolen–Schiffer anomaly was also obtained.

Bibliography

[1] Abashian A and Hafner E M 1958 *Phys. Rev. Lett.* **1** 255

[2] Abegg R *et al* 1986 *Phys. Rev. Lett.* **56** 2571

[3] Abegg R *et al* 1989 *Phys. Rev.* D **39** 2464

[4] Baker C A *et al* 2006 *Phys. Rev. Lett.* **97** 131801

[5] Baker C A *et al* 2007 *Phys. Rev. Lett.* **98** 149102

[6] Balzer R, Henneck R, Jacquemart Ch, Lang J, Simonius M, Haeberli W, Weddigen Ch, Reichart W and Jaccard S 1980 *Phys. Rev. Lett.* **44** 699

[7] Balzer R *et al* 1984 *Phys. Rev.* C **30** 1409

[8] Baumgartner E, Conzett H E, Shield E and Slobodrian R J 1966 *Phys. Rev. Lett.* **16** 105

[9] Berdoz A R *et al* (E497 TRIUMF Collaboration) 2001 *Phys. Rev. Lett.* **87** 272301

[10] Blatt J M and Jackson J D 1950 *Rev. Mod. Phys.* **22** 77

[11] Blanke E, Driller H, Glöckle W, Genz G, Richter A and Schrieder G 1983 *Phys. Rev. Lett.* **51** 355

[12] Bodanski D, Eccles S F, Farwell G W, Rickey M E and Robison P C 1959 *Phys. Rev. Lett.* **2** 101

[13] Bodanski D, Braithwaite W J, Shreve D C, Storm D W and Weitkamp W G 1966 *Phys. Rev. Lett.* **17** 589

[14] Brickwedde F A, Dunning J R, Hoge H J and Manley J H 1938 *Phys. Rev.* **54** 266

[15] Conzett H E 1993 *Phys. Rev.* C **48** 423

[16] Conzett H E 1994 *Rep. Prog. Phys.* **57** 1

[17] Desplanques B, Donoghue J F and Holstein B R 1980 *Ann. Rev. Phys. NY* **124** 449

[18] Engels R *et al* (JEDI Collaboration) http://www2.fz-juelich.de/ikp/jedi/documents/proposals.shmtl

[19] Eversheim D *et al* 1991 *Phys. Lett.* B **256** 11

[20] Fermi E and Marshall L 1947 *Phys. Rev.* **71** 666

[21] Frauenfelder H and Henley E M 1986 *Nuclear and Particle Physics A: Background and Symmetries* (*Lecture Notes and Supplements in Physics*) (Reading: Benjamin/Cummings)

[22] Gonzáles-Trotter D E *et al* 1999 *Phys. Rev. Lett.* **83** 3788

[23] Haddock R P, Salter R M, Zeller M, Czirr J B and Nygren D R 1965 *Phys. Rev. Lett.* **14** 318

[24] Halpern J, Estermann E, Simpson O C and Stern O 1937 *Phys. Rev.* **52** 142

[25] Handler R, Wright S C, Pondrom L, Limon P, Olsen S and Kloeppel P 1967 *Phys. Rev. Lett.* **19** 933

[26] Harney H L, Richter A and Weidenmüller H A 1986 *Rev. Mod. Phys.* **58** 607

[27] Harney H L, Hüpper A and Richter A 1990 *Nucl. Phys.* A **518** 35

[28] Haxton W C and Holstein B R 2013 arXiv: 1303.4132v2

[29] Henley E M and Jacobsohn B A 1959 *Phys. Rev.* **113** 225

[30] Henley E M and Miller G A 1979 *Mesons in Nuclei* ed M Rho and H D Wilkinson (Amsterdam: North-Holland) p 415

[31] Hillman P, Johansson A and Tibell G 1958 *Phys. Rev.* **110** 1218

[32] Holstein B R 2009 *Eur. Phys. J.* A **41** 279
[33] Huhn V, Wätzold L, Weber Ch, Siepe A, von Witsch W, Witała H and Glöckle W 2000 *Phys. Rev. Lett.* **85** 1190
[34] Kistryn S *et al* 1987 *Phys. Rev. Lett.* **58** 1616
[35] Klein G and Schieck H Paetz gen. 1974 *Nucl. Phys.* A **219** 422
[36] Knecht D J, Messelt S, Berners E D and Northcliffe L C 1958 *Phys. Rev.* **114** 550
[37] Lockyer N *et al* 1984 *Phys. Rev.* D **30** 1409
[38] McKibben J L, Lawrence G P and Ohlsen G G 1968 *Phys. Rev. Lett.* **20** 1180
[39] Melkonian E 1949 *Phys. Rev.* **76** 1744
[40] Miller G A, Opper A K and Stephenson E J 2006 *Ann. Rev. Nucl. Part. Sci.* **56** 253
[41] Moldauer P A 1968 *Phys. Rev.* **165** 1136
[42] Nagle E D, Bowman J D, Hoffman C, McKibben J, Mischke R, Potter J M, Frauenfelder H and Sorensen L 1979 *AIP Conf. Ser.* **51** 224
[43] Oxley C L, Cartwright W F, Rouvina J, Baskir E, Klein D, Ring J and Skillman W 1953 *Phys. Rev.* **91** 419
[44] Schieck H Paetz gen. 2014 *Nuclear Reactions—An Introduction* (*Lecture Notes in Physics* vol 882) (Heidelberg: Springer)
[45] Rainwater J and Havens Jr W W 1946 *Phys. Rev.* **70** 136
[46] Ryan J W 1964 *Phys. Rev. Lett.* **12** 564
[47] Ryan J W 1967 *Phys. Rev.* **130** 1554
[48] Sprung D W L 1961 *Phys. Rev.* **121** 925
[49] Stephenson E J *et al* 2003 *Phys. Rev. Lett.* **91** 142303
[50] Sutton R B, Hall T, Anderson E E, Bridge H S, de Wire J W, Lavatelli L S, Long E A, Snyder T and Williams R W 1947 *Phys. Rev.* **72** 1147
[51] Tanner N 1957 *Phys. Rev.* **107** 1203
[52] von Witsch W, Richter A and von Brentano P 1968 *Phys. Rev.* **169** 923
[53] Weitkamp W G, Storm D W, Shreve D C, Braithwaite W C and Bodanski D 1968 *Phys. Rev.* **165** 1233
[54] Worthington H R, McGruer J N and Findley D E 1952 *Phys. Rev.* **90** 899
[55] Yuan V, Frauenfelder H, Harper R W, Bowman J D, Carlini R, MacArthur D W, Mischke R E, Nagle D E, Talaga R L and McDonald A B 1986 *Phys. Rev. Lett.* **57** 1680
[56] Zeitnitz B, Maschuw R and Suhr P 1969 *Phys. Lett.* B **28** 420
[57] Zhao J *et al* 1998 *Phys. Rev.* C **57** 2126

Chapter 13

Scattering of identical nuclei, exchange symmetry and molecular resonances

In quantum mechanics identical particles are indistinguishable. In a reaction between identical nuclei this applies to forward-scattered nuclei and backward-scattered recoil nuclei. These particles must interfere independent of the reaction mechanisms and the type of interaction, i.e. classical or semi-classical approaches to describe such reactions are unsuitable in principle.

13.1 The first observation of interference in the scattering of identical nuclei

To best see the quantum interference of identical particles it is useful to stay below the Coulomb barrier (*sub-Coulomb scattering*), i.e. where nuclear effects can be kept small. Then one has to deal with calculable (Rutherford) amplitudes only. In addition to the forward scattering Rutherford cross section there is a corresponding recoil Rutherford term plus an interference term between both. In the scattering of identical particles a detector at the center-of-mass (c.m.) angle θ is unable to distinguish whether it registers forward-scattered ejectiles under θ or, under the angle $\pi - \theta$, backward-emitted recoils. This is shown in figure 13.1. The formal scattering theory (see below) shows that the angular distributions must be symmetric around $\pi/2$ and therefore must be described by even-order Legendre polynomials. Quantum-mechanically, in addition, it is to be expected that the forward- and backward-scattered particle waves interfere. In this case no classical description of the scattering process is possible. In addition, the details of the interference depend on the spin structure of the interacting particles: identical bosons behave differently from identical fermions and when the particles have spin $\neq 0$ (i.e. always for fermions) the spin states must be coupled and superimposed in the cross section with

doi:10.1088/978-0-7503-1173-1ch13

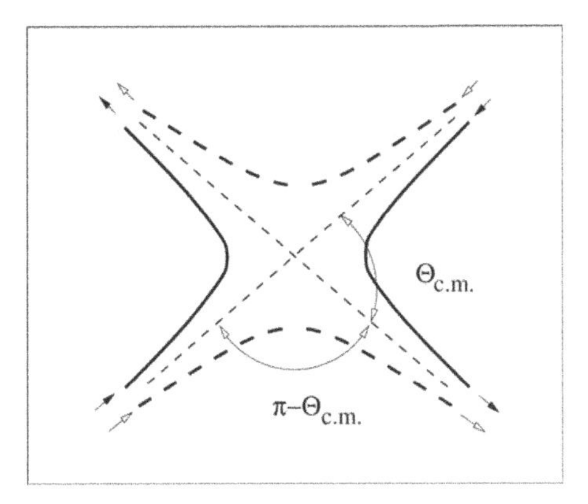

Figure 13.1. Trajectories of identical particles in the c.m. system.

their spin multiplicities as weighting factors. The following examples, which can be tested experimentally, will explain this.

13.1.1 Identical bosons with spin $I = 0$

Here

$$\left[\mathrm{d}\sigma/\mathrm{d}\Omega(\theta)\right]_{\mathrm{B}} = \left|f_1(\theta) + f_2(\pi - \theta)\right|^2. \tag{13.1}$$

13.1.2 Identical fermions with spin $I = 1/2$

For the fermions the spin singlet cross section

$$\left[\mathrm{d}\sigma/\mathrm{d}\Omega(\theta)\right]_{\mathrm{s}} = \left|f_1(\theta) - f_2(\pi - \theta)\right|^2 \tag{13.2}$$

and the triplet cross section

$$\left[\mathrm{d}\sigma/\mathrm{d}\Omega(\theta)\right]_{\mathrm{t}} = \left|f_1(\theta) + f_2(\pi - \theta)\right|^2 \tag{13.3}$$

in the total (integrated) cross section must be added *incoherently*, each weighted with their spin multiplicities:

$$\left[\mathrm{d}\sigma/\mathrm{d}\Omega(\theta)\right]_{\mathrm{F}} = \tfrac{1}{4}\left|f_1(\theta) + f_2(\pi - \theta)\right|^2 + \tfrac{3}{4}\left|f_1(\theta) - f_2(\pi - \theta)\right|^2. \tag{13.4}$$

In these two cases the interference has opposite signs which, e.g. at $\theta = \pi/2$, has the consequence that in the case of two bosons there is an interference maximum, for fermions a minimum. Under the special assumption that there is no spin–spin force acting ($f_s = f_t = f$), and with $f(\theta) = f(\pi - \theta)$ one obtains for identical fermions a decrease, for identical bosons an increase each by a factor of 2 compared to the classical cross section.

For pure (sub-)Coulomb scattering (meaning: Coulomb scattering at energies sufficiently below the Coulomb barrier) of identical particles the scattering amplitudes can be calculated explicitly (i.e. also summed over partial waves) since we deal with the Rutherford amplitude known from scattering theory, see section 2.1.3:

$$\left(\frac{d\sigma}{d\Omega}\right)_{\text{Coul}} = \left(\frac{Z^2e^2}{4E_\infty}\right)^2 \left[\frac{1}{\sin^4\frac{\theta}{2}} + \frac{1}{\cos^4\frac{\theta}{2}} + \frac{2(-1)^{2s}\cos\left(\eta_S \ln\tan^2\frac{\theta}{2}\right)}{(2s+1)\sin^2\frac{\theta}{2}\cos^2\frac{\theta}{2}}\right]. \tag{13.5}$$

In addition to the forward scattering Rutherford cross section there is a corresponding recoil Rutherford term plus an interference term between both. Figure 13.2 shows this behavior, which is analogous to that of light in Young's double-slit experiment, but additionally shows the influence of spin and statistics. Two fermions

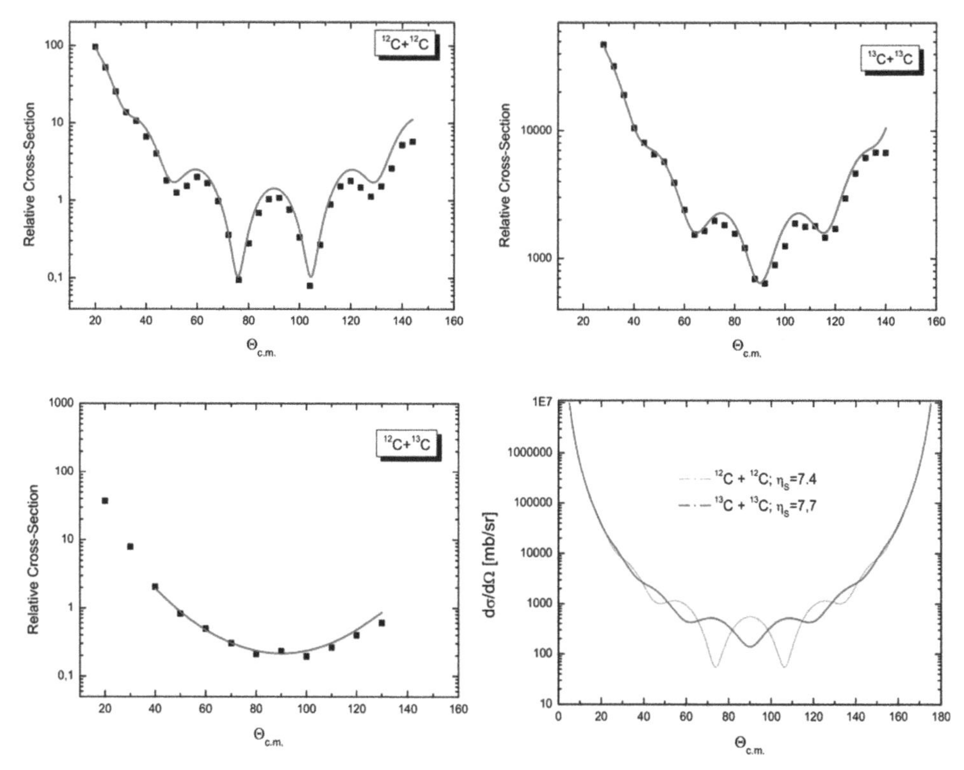

Figure 13.2. Experimental c.m. angular distributions of Coulomb scattering of two identical bosons (^{12}C) and fermions (^{13}C) as well as of two non-identical particles of nearly equal masses and theoretical cross sections at $E_{\text{lab}} = 7$ MeV. The angular distribution for the non-identical particles is obtained when the spectra of the forward- and backward-scattered particles cannot be separated by the detector, which is the case for (nearly) equal masses. Otherwise one would obtain a typical Rutherford distribution for the forward-scattered particle and a distribution reflected about 90° for the recoil particle. The data were measured by students of an advanced laboratory course at IKP Cologne in 2003.

show destructive interference at $\pi/2$ whereas two bosons interfere constructively. In fact the experiments could be used to determine the spins of the particles involved from the 'amplitude' of the interference pattern. Above the Coulomb barrier, additional terms including interference terms arise from the hadronic interaction. A special example is low-energy proton–proton scattering in which, for S-waves, one nuclear phase shift δ_0 must be considered for which, e.g. a 'nuclear' scattering length $a_{\rm pp}$ may be obtained (see section 12.3.2).

13.2 Studies of heavy-ion reactions and intermediate structure

The first and key experiments designed specifically to study the scattering of identical bosons and at the same time details of the heavy-ion interactions and structure were performed by Bromley *et al* [4, 5]. Scattering of identical spin-zero nuclei was investigated. Larger tandem Van de Graaff accelerators had become available as well as compact solid-state detectors. Both are especially suited to studying heavy-ion reactions, the accelerators because of the possible high energies for heavy ions due to the multiple charge states of the ion beams, excellent energy definition and stability, and ease of changing energy and targets, the latter because of the possibility of using many small detectors providing good angular resolution and relatively thin depletion layers. Thus, measurements of the highly structured cross sections, both in angle and energy, were facilitated. Figure 13.3 shows the scattering

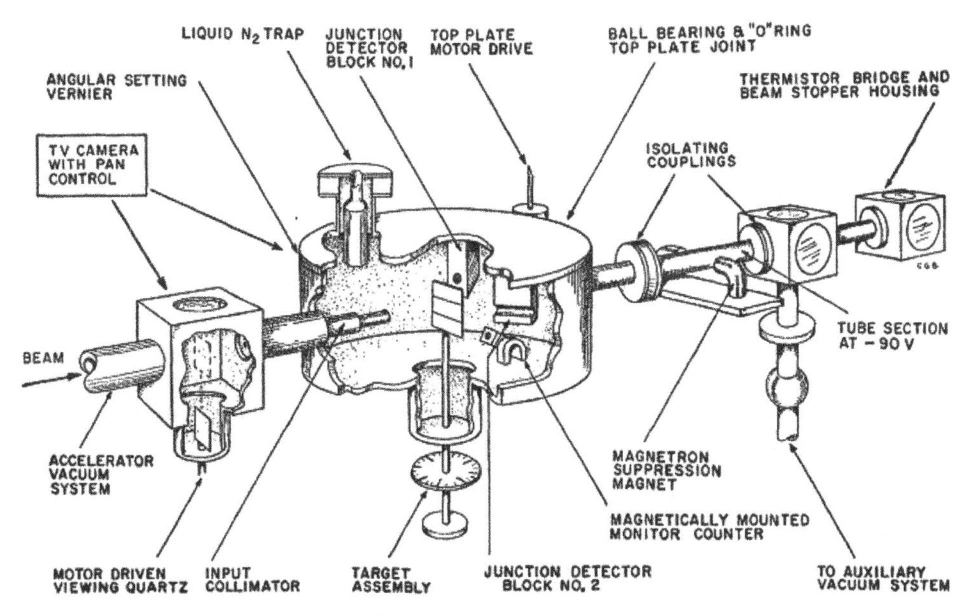

Figure 13.3. The scattering chamber set-up of [5] is shown. It is typical for tandem Van de Graaff experiments with charged particles. Important are a well-focused or collimated beam, thin foil targets, well-defined solid angles at the detectors, precise angle definition of the detector slits and a Faraday cup at the exit for beam charge calibration. For heavy-ion experiments, however, this cannot be used because the charge state equilibrium of the incident beam at the reaction is unknown. Therefore, a cross-section calibration such as using a heavy target (e.g. gold) and a calculable Rutherford cross section has to be used for normalization. Reproduced with permission from [5]. Copyright 1961 American Physical Society.

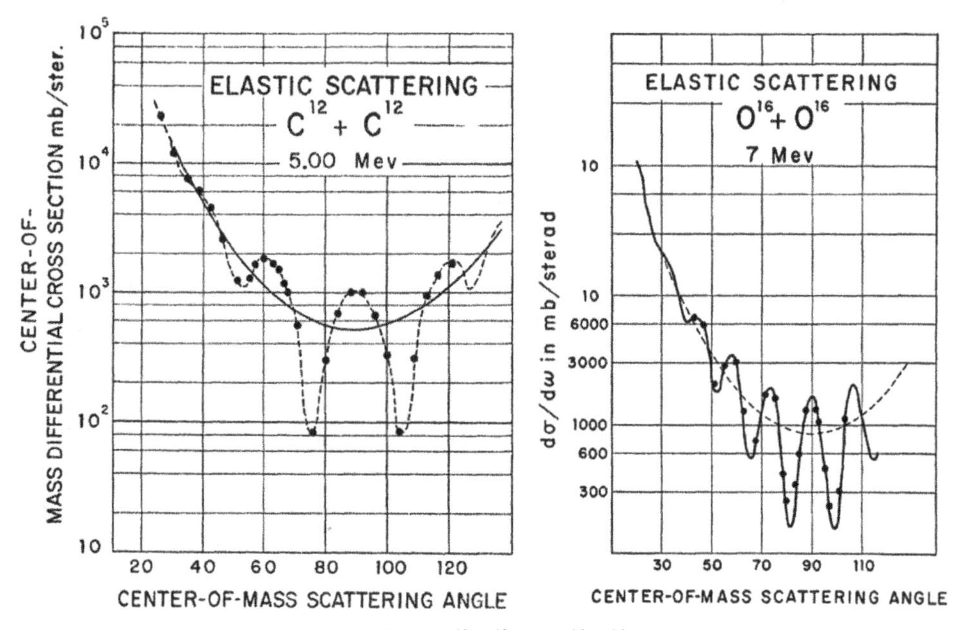

Figure 13.4. Cross-section angular distributions of ^{12}C–^{12}C and ^{16}O–^{16}O elastic scattering. Reproduced with permission from [5]. Copyright 1961 American Physical Society.

chamber set-up used in [5]. Figure 13.4 shows a selection of the angular distributions for ^{12}C–^{12}C and ^{16}O–^{16}O elastic scattering at energies below the Coulomb barriers such that the Mott cross section applies. The interference pattern for identical boson scattering is evident. One of the earliest measurements of elastic scattering between identical fermions (^{13}C–^{13}C) was performed by Voit *et al* [9], see also figure 13.2 for details of the spin and statistics. Figure 13.5 compares excitation functions for ^{12}C–^{12}C and ^{16}O–^{16}O elastic scattering. It is remarkable that the excitation functions for the two systems are quite different, i.e. relatively smooth for the ^{16}O case, but with (quasi-)periodic structures for the ^{12}C scattering (*intermediate* or *gross* structures with superimposed *fine* structure). These would be typical signatures for a *doorway* phenomenon. In fact, these oscillations have been interpreted as rotational states of nuclear molecules which partly decay into CN fine-structure states. The excitation functions follow the Mott cross section but with increasing energy there is a relatively sharp onset of strong absorption (leading to compound-nucleus formation). The r dependence of the combined potential of the Coulomb potential, an absorptive (attractive) nuclear potential and the orbital angular-momentum barrier at higher ℓ could form a shallow minimum with rotational molecular states as shown in figure 13.6. In contrast to these early measurements more detailed studies, also at higher energies, exhibit (quasi-)periodic structures, partly with marked fine-structure oscillations, also in other systems such as ^{16}O–^{16}O, ^{12}C–^{16}O, ^{14}C–^{14}C, ^{18}O–^{16}O, and ^{18}O–^{18}O, see [16]. Three cases are shown in figure 13.7. Later measurements showed intermediate structures in excitation functions also in several

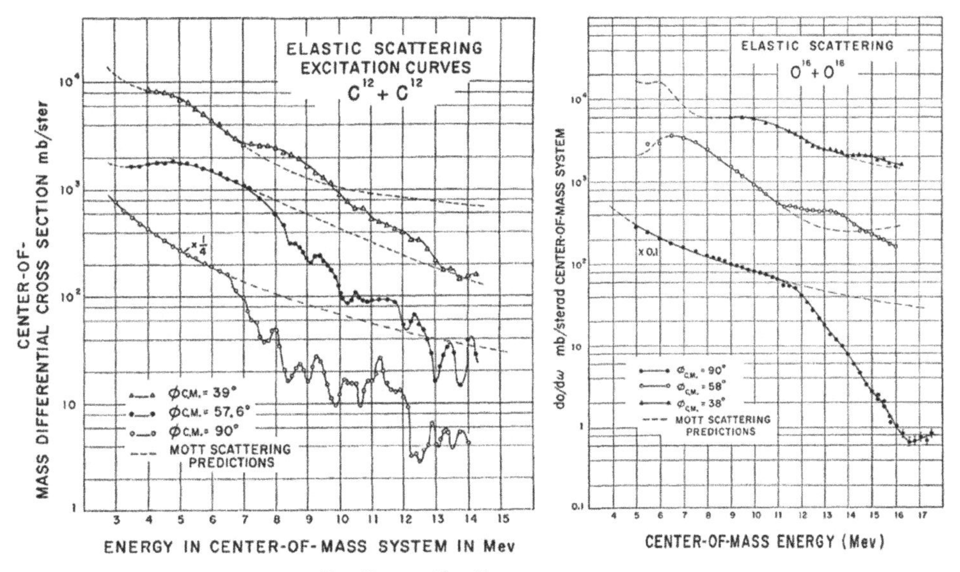

Figure 13.5. Excitation functions of ^{12}C–^{12}C and ^{16}O–^{16}O elastic scattering. Reproduced with permission from [5]. Copyright 1961 American Physical Society.

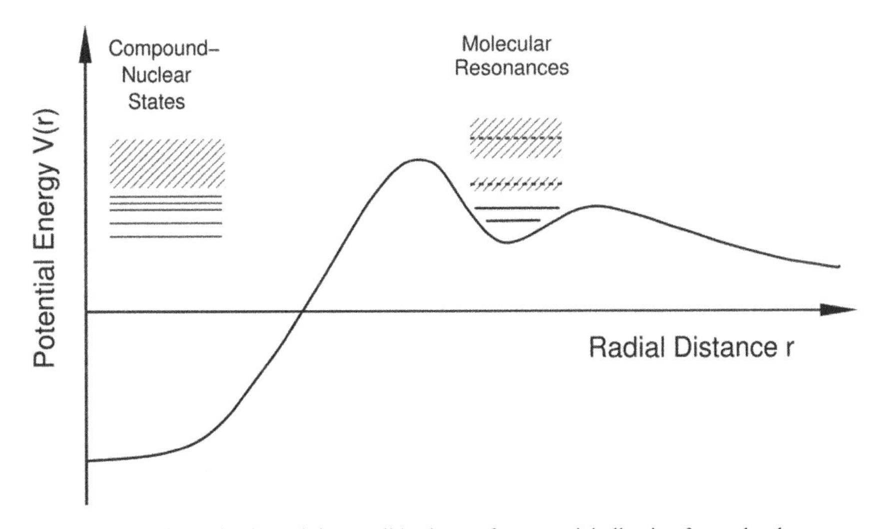

Figure 13.6. Schematic view of the possible shape of a potential allowing for molecular states.

combinations of nuclei, even in ^{16}O–^{16}O, but also between non-identical nuclei, ^{12}C–^{16}O, ^{18}O–^{16}O and ^{18}O–^{18}O, partly with superimposed (CN) fine structure. An example for a precise measurement of the fine structures is shown in figure 13.8. For a detailed discussion of nuclear molecular states and attempts to describe them, e.g. in the framework of the optical model with shallow or deep optical potentials, see [2]. For general reading on heavy-ion nuclear physics see the selected references [1, 3, 6, 8, 10–13, 15].

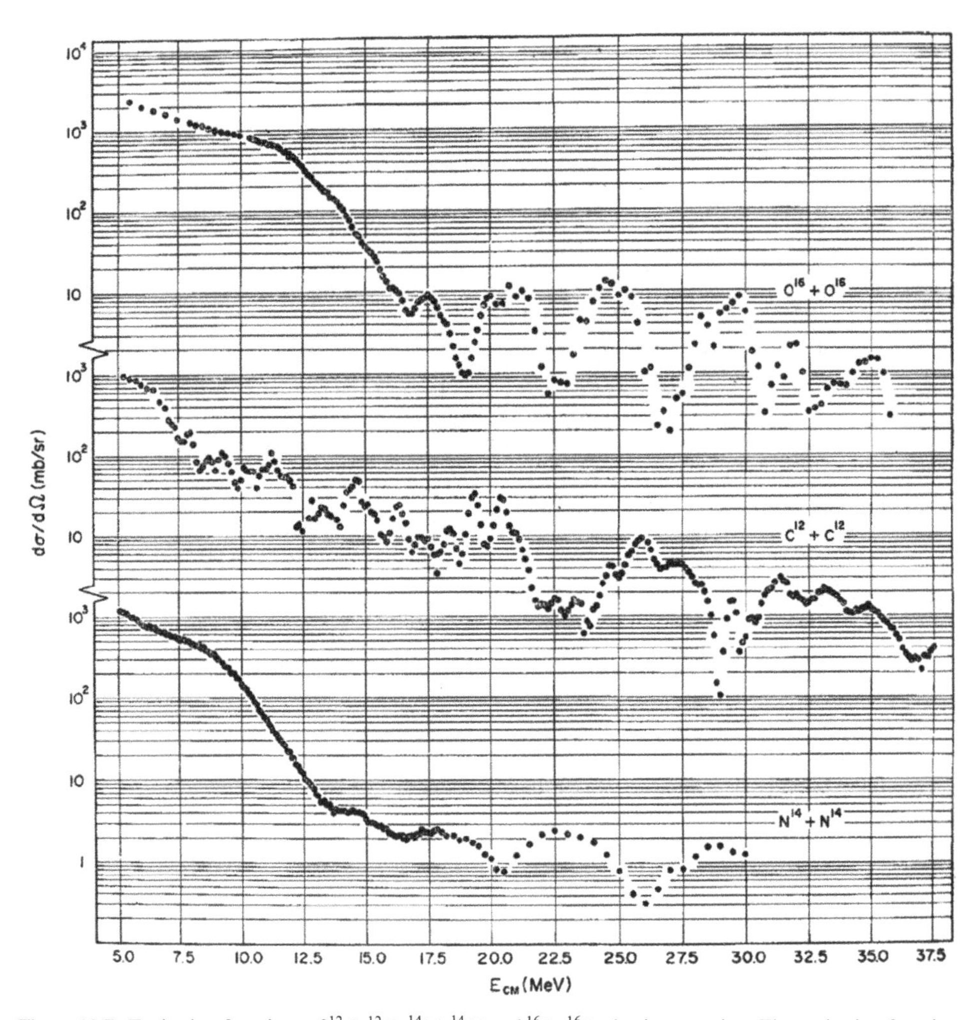

Figure 13.7. Excitation functions of ^{12}C–^{12}C, ^{14}N–^{14}N and ^{16}O–^{16}O elastic scattering. The excitation functions show indications of fine structure superimposed on intermediate (molecular resonance) structures. Reproduced with permission from [14]. Copyright 1973 Springer.

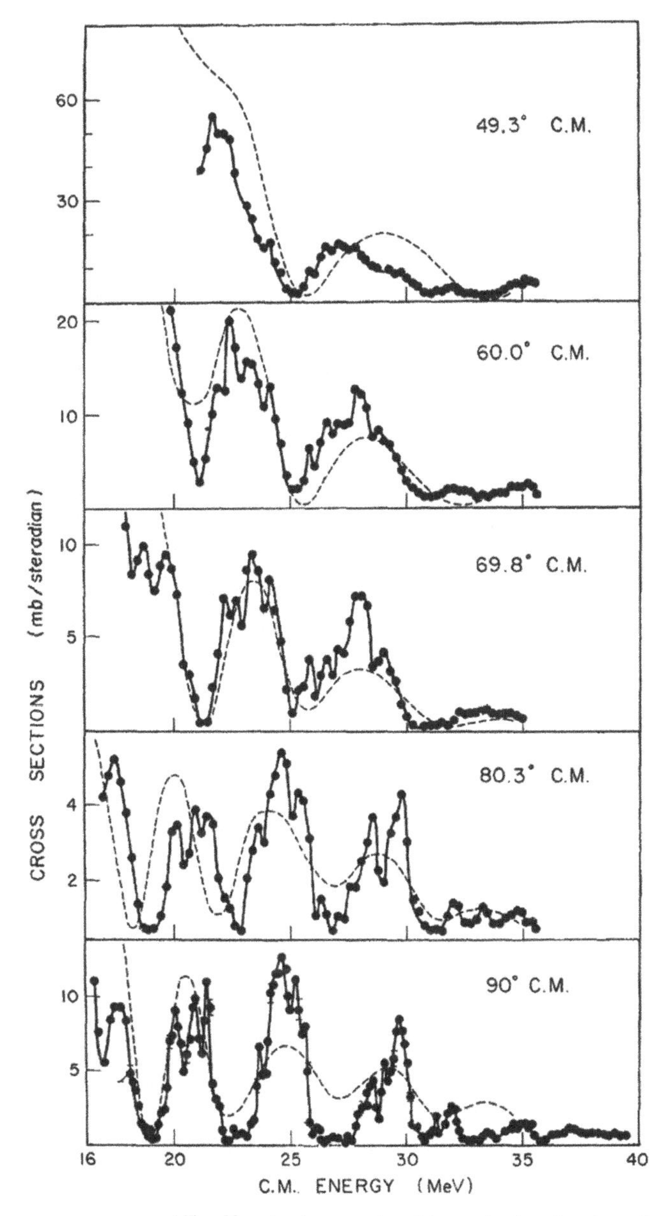

Figure 13.8. Excitation functions of ^{16}O–^{16}O elastic scattering. The excitation functions show a well-resolved fine structure superimposed on intermediate (molecular resonance) structures. The dashed line is from an optical-model calculation with an extremely shallow real potential. Reproduced with permission from [16]. Copyright 1967 American Physical Society.

Bibliography

[1] Bass R 1980 *Nuclear Reactions with Heavy Ions (Texts and Monographs in Physics)* (Berlin: Springer)

[2] Betts R R and Wuosmaa A H 1997 *Rep. Prog. Phys.* **60** 819

[3] Bock R (ed) 1981 *Heavy-Ion Collisions* vol 1–3 (Amsterdam: North-Holland)

[4] Bromley D A, Kuehner J A and Almqvist E 1960 *Phys. Rev. Lett.* **4** 365

[5] Bromley D A, Kuehner J A and Almqvist E 1961 *Phys. Rev.* **123** 868

[6] Bromley A (ed) 1985 *Treatise on Heavy-Ion Science* vol 1–8 (New York: Plenum)

[7] Frahn W E 1972 *Ann. Phys. NY* **72** 524

[8] Goldberger M L and Watson K M 1964 *Collision Theory* (New York: Wiley)

[9] Helb H-D, Dück P, Hartmann G, Ischenko G, Siller F and Voit H 1973 *Nucl. Phys.* A **206** 385

[10] Joachain C 1983 *Quantum Collision Theory* 3rd edn (Amsterdam: North-Holland)

[11] Mott N F 1930 *Proc. R. Soc.* A **126** 259

[12] Mott N F and Massey H S W 1965 *The Theory of Atomic Collisions* (Oxford: Clarendon)

[13] Nörenberg W and Weidenmüller H A 1976 *Introduction to the Theory of Heavy Ion Collisions (Lecture Notes in Physics)* vol 51 (Heidelberg: Springer)

[14] Reilly W, Wieland R, Gobbi A, Sachs M W and Bromley D A 1973 *Nuovo Cimento* A **13** 897

[15] Satchler G R 1990 *Introduction to Nuclear Reactions* 2nd edn (London: McMillan)

[16] Siemssen R H, Maher J V, Weidinger A and Bromley D A 1967 *Phys. Rev. Lett.* **19** 369

Chapter 14

Nuclear fission and nuclear energy

Nuclear fission—i.e. the disintegration of a heavy nucleus into two (sometimes three) lighter nuclei of roughly the mass and charge numbers of the original—can occur spontaneously in some heavy nuclei, but there are many instances of fission induced by nuclear reactions. All kinds of combinations of projectiles and heavy target nuclei may undergo fission. Nuclei, excited to very high (rotational) spin states could also fission.

After the discovery of the neutron by Chadwick in 1932 many groups started to study the interaction between neutrons and nuclei. The property of neutrality of the neutron made it very attractive as a probe of nuclei and nuclear reactions especially at low energies where no Coulomb barrier hindered the reactions. Even with the simple method of using radioactive sources such as Ra–Be and hydrogen-containing moderators (such as water or paraffin) many new results could be obtained. The practical use of accelerators after 1932 opened additional possibilities as did much later the use of nuclear reactors with high neutron fluxes.

In the 1930s Enrico Fermi and his collaborators tried to produce transuranic nuclei by adding a neutron to a known nucleus. They succeeded in the case of $^{239}_{93}Np$ via the neutron capture of $^{238}_{92}U$ and subsequent β decay. At the suggestion of Lise Meitner, Otto Hahn's group started performing similar experiments with the goal of producing new elements. Only in 1938 did Otto Hahn and Friedrich Strassmann (after Meitner had been forced to emigrate to Sweden) take the additionally produced activities, which looked like Ba and other medium-weight nuclei, seriously and identified these medium-weight products as *fission products*.

The nuclear chemists Hahn and Strassmann published a series of papers in 1938/39, in which they clearly proved—using chemical means—that after the irradiation of uranium (and thorium) with neutrons not only transuranium elements were created, as was generally assumed, see [10], but also that medium-heavy isotopes of barium (and lanthanum and cerium) unambiguously appeared [12].

doi:10.1088/978-0-7503-1173-1ch14

The experimental set-up was remarkably simple: slow neutrons were produced by a radium–beryllium source, followed by a paraffin moderator, and used to irradiate, e.g., a uranium compound such as uranyl nitrate with slowed-down neutrons. The captured neutrons left the reaction products in excited states and the decay activity was measured using a Geiger counter. A set-up which collects the different pieces of apparatus is exhibited at the Deutsches Museum, Munich, see figure 14.1.

Radium is chemically homologous to barium and thus both could appear together in chemical separations. Therefore, in the beginning, Hahn *et al* assumed that they had produced new isotopes of radium from α decays of transuranium nuclei. With more refined chemical methods ('fractionated crystallization') barium and radium could be separated and no trace of enrichment of radium (with its known half-life) could be found. Thus, they could only conclude that they had produced barium isotopes. It is interesting to read in the original paper how difficult this conclusion was for them from the point of view of physics, but as chemists they could not evade it. In translation: 'As chemists in principle we should rename the above scheme and replace the symbols or Ra, Ac, and Th by Ba, La, and Ce. As 'nuclear chemists' who are in a certain way close to physics we cannot yet make up our mind for this jump. Perhaps a number of strange accidents could still have simulated our results' [11]. But only a little later [12] they wrote in the conclusion, 'The generation of barium isotopes from uranium has finally been proven', and also isotopes of Sr and Y as fragments, similarly for thorium.

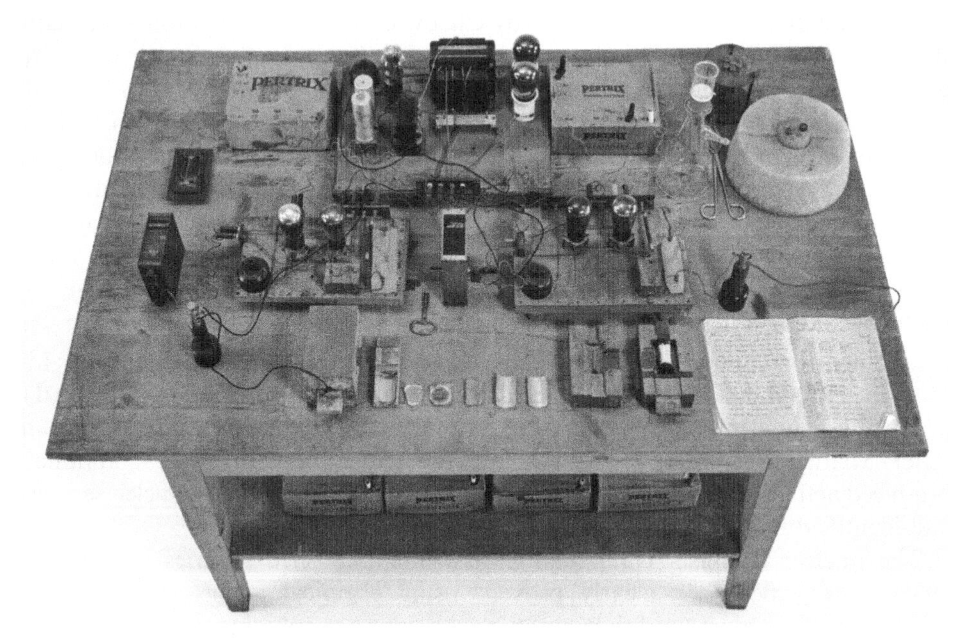

Figure 14.1. Museum exhibit at the Deutsches Museum, Munich: the famous Hahn–Meitner–Strassmann desk with the collected pieces of experimental equipment and Hahn's laboratory logbook. It is likely that the equipment was mostly used by Meitner, as she was the physicist in a group of nuclear chemists. The pieces of equipment were actually used in different rooms. Photo and Copyright Deutsches Museum, München.

This hesitation shows how improbable the only possible conclusion of fission of heavy nuclei into approximately equal debris appeared. It was accepted knowledge that nuclei could not fission. It is, however, difficult to understand that Fermi and his group at Rome did not conceive of fission during their long series of experiments with neutrons produced from different sources such as Ra–Be, Po–Be, etc, on many different elements, among them uranium and thorium. The results were published in ten short communications (letters) in Italian in the journal *Ricerca Scientifica* of the National Research Council and summarized in [4–7]. The group was so fixated on producing 'new' (transuranic) nuclei that even the idea of possible fission put forward by Ida Noddack as early as 1934 [17] was ignored (for a good account of this see e.g. [19, 20]). Part of this was energetic considerations, the unfounded belief that only α particles could leave the nucleus and the lack of a plausible model (however, around that time the liquid-drop model appeared just in time to provide an explanation).

It took almost five years before the fission process was established. The experiments were confirmed very quickly in many places and the possibility of fission was theoretically explained with the analogy of nuclei as liquid drops, first by Meitner and Frisch [16], then by Bohr and Wheeler [1]. The basic idea of nuclei as liquid drops had been developed by George Gamow in 1928, an achievement partly ignored by a number of later authors [21, 25]. The first *physical*, as compared to *chemical*, proof of the fission process was published by Otto Frisch in [9]. He (at Copenhagen) used an ionization chamber lined with uranium irradiated with neutrons produced by a Ra–Be α source with and without a paraffin moderator. In the first case the number of fission events was approximately twice that of the unmoderated case. When using thorium as the 'target' no dependence on moderation was observed. The medium-mass fission fragments with their high energies and high charge states (estimated to about 20) could travel a few mm into the ionization volume and caused high pulses in the connected counter whereas 'transuranes' would not be able to leave the uranium lining.

Two important consequences of the phenomenon of nuclear fission were also discussed very soon after the discovery.

- The fission process liberates a large amount of energy ($\approx$200 MeV per fission), immediately inciting the idea of technical (and military?) uses.
- The fission products are nuclei with high neutron excess. This neutron excess can be disposed of in two ways, the slower β decay process (partly with delayed neutron emission, a process important for reactor control) or prompt neutron emission. Proof of fission neutrons was first found in 1939 by Dodé *et al* [3] and their number per fission was first determined by von Halban *et al* [13] to be

$$\nu = 3.5 \pm 0.7. \tag{14.1}$$

So, the emission of fast neutrons opened up the possibility of a *chain reaction* that was later implemented in nuclear reactors and nuclear weapons. As early as 1933 Leo Szilard had conceived of a chain reaction (not for a fission reaction, but for the $^{9}Be(n, 2n)2^{4}He$ reaction); he applied for two British patents in 1935 and 1936 instead of publishing the idea, probably out of fear of German developments [22, 23]. Many groups immediately grasped the importance of these discoveries and published calculations of the enormous amounts of energy that could be liberated in the fission process, before all research became classified. As an example, an interesting paper on these considerations was published in June 1939 by Flügge [8] (with the title: 'Kann der Energieinhalt der Atomkerne technisch nutzbar gemacht werden?').

The high energy release connected with nuclear fission led to the known *dual-use* consequences of nuclear power reactors for energy production, but also to nuclear explosives of unprecedented destructive power. Reactors for scientific research use neutrons for many applications, but are also used to breed plutonium for weapons. In all cases the high radioactivity of fission products and the longevity of transuranium elements cause serious problems, and it is hoped that—again with the help of nuclear transmutation reactions—these can be reduced to manageable levels.

For further reading on the physics of fission, a few selected earlier reviews are [2, 18, 24, 26]. A recent text with many references is [15]. A new article 'On the belated discovery of fission' by Pearson [19] with the subtitle 'A remarkable sequence of missteps, misfortune, and oversights delayed the discovery of nuclear fission …' describes in a very compact way the winding road to fission and gives credit to those involved, such as Fermi, Joliot-Curie, Hahn, Meitner, Strassmann, Frisch, Bohr, Szilard, and in particular (the often unrecognized) Noddack who suggested the possibility of nuclear breakup into comparably sized nuclei as early as 1934, together with pertinent references.

Bibliography

[1] Bohr N and Wheeler J A 1939 *Phys. Rev.* **56** 426

[2] Brack M, Damgaard J, Jensen A S, Pauli H C, Strutinsky V M and Wong C Y 1972 *Rev. Mod. Phys.* **44** 320

[3] Dodé M, von Halban. H, Joliot F and Kowarski L 1939 *C. R. Acad. Sci. Paris* **208** 995

[4] Fermi E 1934 *Nature* **133** 757

[5] Fermi E 1934 *Nature* **133** 898

[6] Fermi E 1934 *Nature* **134** 668

[7] Fermi E, Amaldi E, D'Agostino O, Rasetti F and Segrè E 1934 *Proc. R. Soc.* A **133** 483

[8] Flügge S 1939 *Naturwissenschaften* **27** 402

[9] Frisch O R 1939 *Nature* **143** 276

[10] Hahn O and Strassmann F 1938 *Naturwissenschaften* **26** 756

[11] Hahn O and Strassmann F 1939 *Naturwissenschaften* **27** 11

[12] Hahn O and Strassmann F 1939 *Naturwissenschaften* **27** 89

[13] von Halban H, Joliot F and Kowarski L 1939 *Nature* **143** 470 and 680

[14] Exhibit at Deutsches Museum, Munich

[15] Krappe H J and Pomorski K 2012 *Theory of Nuclear Fission* (*Lecture Notes in Physics*) vol 838 (Heidelberg: Springer)

[16] Meitner L and Frisch O R 1939 *Nature* **143** 239
[17] Noddack I 1934 *Angew. Chem.* **47** 653
[18] Pauli H C 1973 *Phys. Rep.* **7** 35
[19] Pearson J M 2015 *Phys. Today* **68** no. 6, 40
[20] Segrè E 1970 *Enrico Fermi, Physicist* (Chicago: University of Chicago Press)
[21] Stuewer R H 1994 *Perspect. Sci.* **2** 76
[22] Szilard L 1968 *Reminiscences* (*Perspectives in American History* vol 2) (New York: Cambridge University Press) p 94
[23] Feld B T and Weiss Szilard G (ed) 1972 *The Collected Works of Leo Szilard—Scientific Papers* (Cambridge, MA: MIT Press) p 622
[24] Vandenbosch R and Huizenga J R 1973 *Nuclear Fission* (New York: Academic)
[25] von Weizsäcker C F 1937 *Die Atomkerne: Grundlagen und Anwendungen ihrer Theorie* (Leipzig: Akad. Verlagsgesellschaft)
[26] Wilets L 1964 *Theories of Nuclear Fission* (Oxford: Clarendon)

IOP Publishing

Key Nuclear Reaction Experiments
Discoveries and consequences
Hans Paetz gen. Schieck

Chapter 15

The first double scattering and polarization in p–^{4}He and the $(\ell \cdot \mathbf{s})$ force

In 1949 the single-particle shell model of nuclear states was formulated by Haxel, Jensen and Suess [4], and Goeppert-Mayer [3]. It became the basis of nuclear structure models up to today. The essential ingredient in this model is the non-central spin–orbit force term in the nucleon–nucleus potential. Orbitals with a given orbital angular momentum ℓ can be split by the spin–orbit force into doublets corresponding to states with $j = \ell \pm 1/2$. Depending on the sign of this force the sequence of the levels could be 'normal', i.e. (like in atomic physics) the level with the higher j could be raised in energy, the other lowered or vice versa, i.e. 'inverted'. It turned out that the level ordering in nuclei is different from that in atomic states, meaning that the origin of the $(\ell \cdot \mathbf{s})$ force was probably not electromagnetic. The splitting increases with increasing ℓ such that increasingly the spin–orbit force lowers the energy of the lower state so much that in higher orbits it is now in the next-lower harmonic oscillator shell with opposite parity, thus explaining *isomeric* states with longer lifetimes because of larger spin differences and parity change in comparison to the surrounding lower ℓ states. This lowering of states creates the large energy gaps that are characteristic for the shell model.

Until 1949 the level ordering in ^{5}Li* and ^{5}He* was unsettled. The unbound ground and excited states states could be studied as resonances in ^{4}He + p or ^{4}He + n scattering. With the help of phase-shift analyses predictions could be made for the polarization of nucleons scattered from ^{4}He that differed appreciably for the two cases $^2P_{1/2}$ and $^2P_{3/2}$.

For low-energy neutrons polarization effects in the interaction of the neutron's magnetic moment and magnetized material (i.e. the polarized electrons) had been investigated earlier. Theoretical investigations of the effects of an $(\ell \cdot \mathbf{s})$ force (see [9, 10, 12]) on nuclear reactions, e.g. production and measurements of spin polarization, for fast neutrons or protons predicted that

- The scattered particles would acquire spin polarization.
- By time-reversal invariance the scattering, if performed with polarized projectiles, would experience a left–right asymmetry (i.e. an azimuthal dependence of the cross sections): the reaction would show a non-zero analyzing power. Time-reversal invariance states that the polarization produced in a forward reaction is equal to the analyzing power in the corresponding backward reaction, provided the reactions are measured at the same center-of-mass (c.m.) energies and scattering angles. For elastic scattering both are the same (see also [6]).
- Combining both—a first scattering producing a polarization which in turn may be determined by a second scattering of the same kind, a double-scattering experiment—would be unique in showing the existence of the ($\ell \cdot \mathbf{s}$) force.
- At the same time the polarization/analyzing power of the above reactions are predicted to be strongly dependent on the j-value of the split shell-model orbital.

Figures 15.1 and 15.2 show the level schemes for ^{5}Li and ^{5}He as of 1950 and 1952, respectively, i.e. before and after the $A = 5$ nuclei had been investigated in detail, especially in view of the $P_{3/2}$–$P_{1/2}$ splitting and thus the existence of a strong spin–orbit interaction. Heusinkveld and Freier [5] undertook the first double-scattering experiment

$$^4\text{He} + \text{p} \rightarrow {}^5\text{Li} \rightarrow {}^4\text{He} + \text{p}. \qquad (15.1)$$

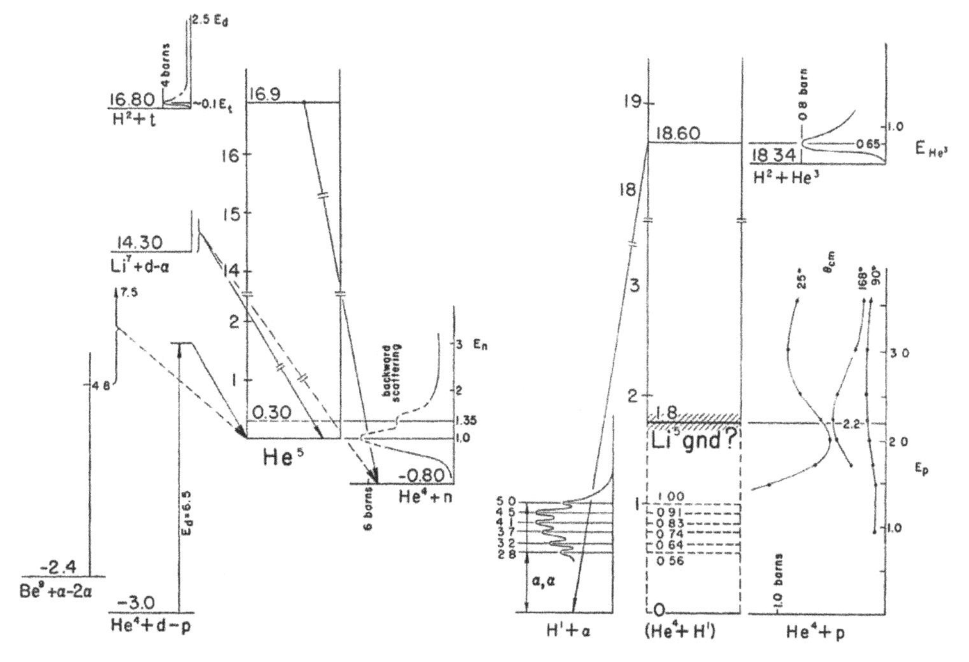

Figure 15.1. Level schemes of ^{5}He and ^{5}Li as of 1950. The size and sequence of the $P_{3/2}$–$P_{1/2}$ splitting of the P state were unclear. Reproduced with permission from [7]. Copyright 1950 American Physical Society.

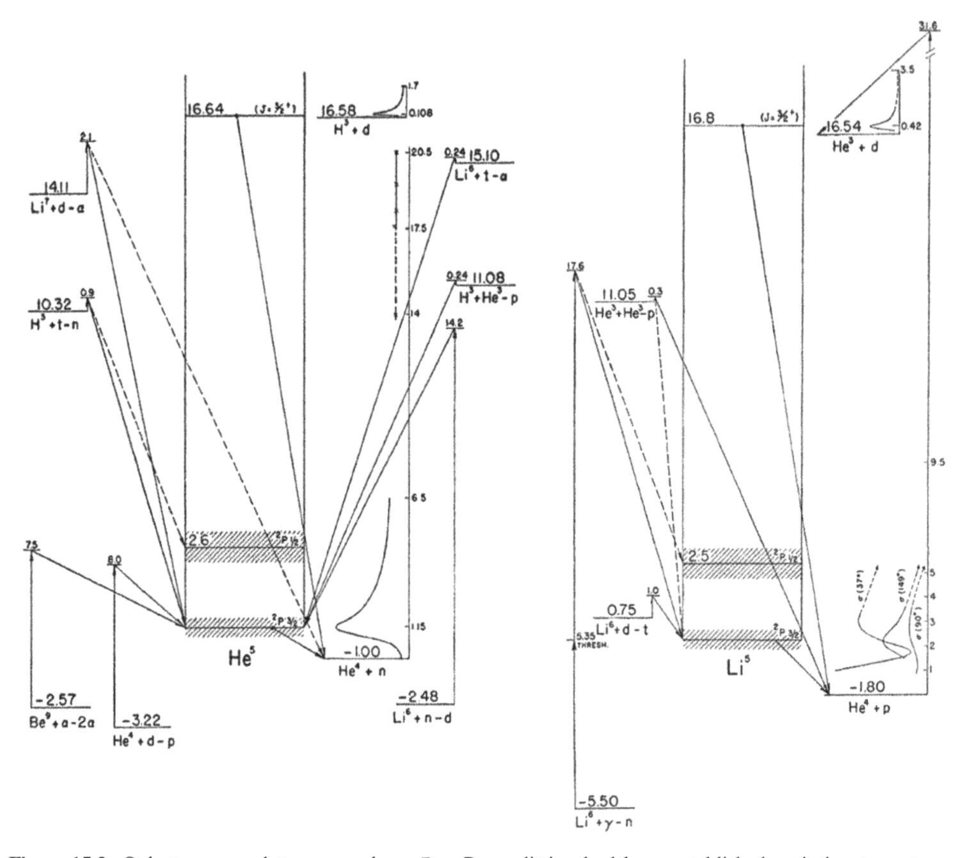

Figure 15.2. Only two years later a very large $P_{3/2}$–$P_{1/2}$ splitting had been established, pointing to a strong spin–orbit force. Aside from small energy corrections this corresponds to our current knowledge, see [11]. Adapted with permission from [8]. Copyright 1952 American Physical Society.

They used photoplates for detecting the doubly scattered protons. Their set-up is depicted in figure 15.3. Figure 15.4 shows the predictions for the polarization for the two cases of a 'regular' or an 'inverted' sequence of the states $P_{1/2}$ and $P_{3/2}$. A phase-shift analysis [1] of cross-section data [2] was used to calculate the polarization of protons scattered from ^{4}He at $\theta_{c.m.} = 90°$ as function of the incident proton laboratory energy. The predictions for the angular distributions yielded marked forward/backward-angle double-scattering cross-section ratios that depended on the level sequence. The results are shown in figure 15.5, a reproduction of the table of results (ratios) from the original. The conclusions from the comparison with the data of this experiment are:

- The P-wave in ^{5}Li (and by isospin arguments also in ^{5}He) is strongly split between $P_{1/2}$ and $P_{3/2}$ which is proof of a strong $(\ell \cdot \mathbf{s})$ force.
- The level sequence is *inverted*, i.e. the origin of the force is not electromagnetic as in the atomic case but a property of the nuclear force itself.
- A scattering experiment $n + {}^4$He or $p + {}^4$He produces highly polarized nucleons. The polarization is measurable in a double-scattering experiment as a left–right asymmetry of the cross section of the second scattering.

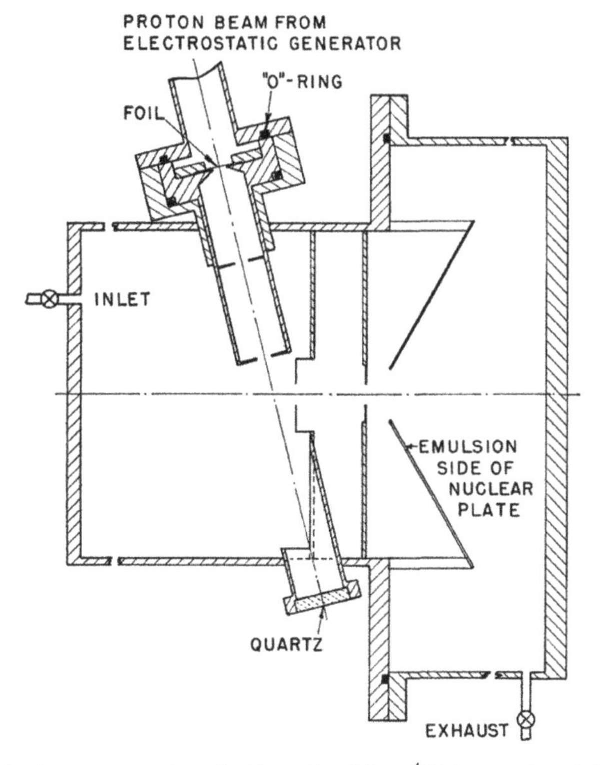

Figure 15.3. Apparatus to measure protons doubly scattered from ^{4}He in a gas target chamber. The protons were registered in the emulsion of photoplates. Reproduced with permission from [5]. Copyright 1950 American Physical Society.

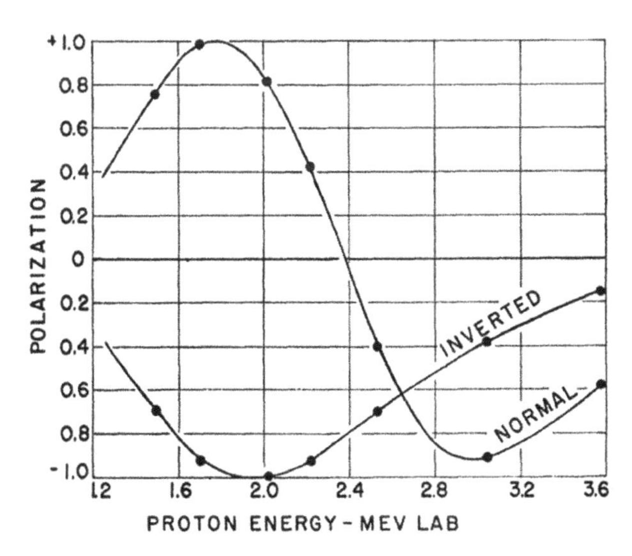

Figure 15.4. Predictions for the proton polarization to be measured in a double-scattering experiment ^{4}He(p, p)^{4}He. Reproduced with permission from [5]. Copyright 1950 American Physical Society.

TABLE I. Ratio of proton tracks in equivalent strips in the forward and backward plates.

	No. of tracks on backward plate	No. of tracks on forward plate	Ratio—backward to forward	Theoretical ratio for ideal ray: Inverted doublet	Theoretical ratio for ideal ray: Normal doublet
3.25 Mev $\frac{1}{8}$-in. slits	364	191	1.9	2.6	1/20.2
3.25 Mev $\frac{1}{4}$-in. slits	220	156	1.4	2.6	1/20.2
3.50 Mev $\frac{1}{8}$-in. slits	61	33	1.85	1.9	1/6.6

Figure 15.5. Reproduction of the table of results of [5] of the double-scattering ratios of ${}^4He(\vec{p}, p){}^4He$. The ratios clearly favor the 'inverted' sequence of levels, i.e. the ${}^2P_{3/2}$ assignment for the ground state and ${}^2P_{1/2}$ for the broad first excited state of ${}^5Li$. Reproduced with permission from [5]. Copyright 1950 American Physical Society.

The $(\ell \cdot \mathbf{s})$ force proved to be the essential ingredient for the shell model of nuclear structure. It is not surprising that nuclear-reaction models such as the optical model (see chapter 10) also require an $(\ell \cdot \mathbf{s})$ term especially for describing polarization observables. The same holds for the nucleon–nucleon interaction to which several spin-dependent forces contribute (see also section 12.3.2).

Bibliography

[1] Critchfield C L and Dodder D C 1949 *Phys. Rev.* **76** 602

[2] Freier G, Lampi E, Sleator W and Williams J H 1949 *Phys. Rev.* **75** 1345

[3] Goeppert-Mayer M 1949 *Phys. Rev.* **75** 1969

[4] Haxel O, Jensen J D H and Suess H E 1949 *Phys. Rev.* **75** 1766

[5] Heusinkveld M and Freier G 1952 *Phys. Rev.* **85** 80

[6] Paetz gen. Schieck H 2012 *Nuclear Physics with Polarized Particles* (*Lecture Notes in Physics* vol 842) (Heidelberg: Springer)

[7] Hornyak W F, Lauritsen T, Morrison P and Fowler W A 1950 *Rev. Mod. Phys.* **22** 291

[8] Ajzenberg-Selove F and Lauritsen T 1952 *Rev. Mod. Phys.* **24** 321

[9] Schwinger J 1946 *Phys. Rev.* **69** 681

[10] Schwinger J 1948 *Phys. Rev.* **73** 407

[11] Tilley D R, Cheves C M, Godwin J L, Hale G M, Hofmann H M, Kelley J H, Sheu C G and Weller H R 2002 *Nucl. Phys.* A **708** 3

[12] Wolfenstein L 1949 *Phys. Rev.* **75** 1664

IOP Publishing

Key Nuclear Reaction Experiments
Discoveries and consequences
Hans Paetz gen. Schieck

Chapter 16

The first nuclear reaction of an accelerated polarized beam from a polarized ion source (Basel)

Usually the quantities most measured in a nuclear reaction are differential or total cross sections. For particles with spin these quantities are—formally—averages over incident-particle spin states and sums over outgoing states or (for the total cross section) also sums over scattering angles (i.e. over orbital angular momenta). In this way much information about the details of the interaction may be lost that might be hidden in the transition amplitudes between the spin-substates, especially regarding the spin dependence of the nuclear interaction, which has, in addition to a central spin-independent term, several spin-dependent contributions such as $(\mathbf{L} \cdot \mathbf{S})$ (see also chapter 15), spin–spin and tensor force contributions.

Up to 1960 the only way to study the spin dependence of nuclear reactions was to produce spin-polarized particles in nuclear reactions with such an $(\ell \cdot \mathbf{s})$ or $(\mathbf{L} \cdot \mathbf{S})$ force acting and/or to measure the polarization of the outgoing particles with such a reaction. The term double scattering for these experiments implies the difficulties involved, such as the dependence on the energy and angle properties of these reactions and the very small event rates.

The 'art' of separating spin states began with the famous *Stern–Gerlach* experiment [6] in which a silver atomic beam was spatially split according to the two values of the atomic (electron) $J = 1/2$ substates in an inhomogeneous magnetic field due to the force $\mathbf{F} = \boldsymbol{\mu}_e \vec{\nabla} |\mathbf{B}|$ on the associated magnetic moment of the atomic electron. This magnetic moment (*g-factor*), the *Bohr magneton* is large enough to effectuate a complete separation of the two partial beams with opposite polarizations even in the presence of thermal broadening. With nucleon/nuclear spins [4, 7] with their associated much (about 2000 times) smaller magnetic moments (nuclear magneton

doi:10.1088/978-0-7503-1173-1ch16

versus Bohr magneton), first measured by Otto Frisch and Otto Stern for the proton [5], such a clean separation is virtually impossible, especially in view of the thermal spread of the atoms' velocities in a beam. However, making use of the hyperfine interaction between nuclear and electron spins (and magnetic moments) and the *Zeeman effect* is the method of choice to separate nuclear spin states together with the atomic fine-structure states.

Starting in 1956 the idea to use these atomic properties for a Stern–Gerlach type separation of spin states was first formulated and realized by Clausnitzer, Fleischmann and Schopper [2, 3] at Erlangen. The electronic polarization of atoms is transferred to the nuclei by the hyperfine interaction. The atomic beam of hydrogen or deuterium produced by a radio frequency (RF) dissociator from H_2 or D_2 gas is fed through a beam-forming device (collimator or nozzle) through the Stern–Gerlach magnets into an ionizer where by electron impact ionization polarized protons or deuterons can be extracted and injected into an accelerator. The first sources used an ionizer in a weak magnetic field producing nuclear spin polarization directly, but not with maximum polarization values.

The first nuclear reaction initiated by a polarized and accelerated beam from such a source was the $^3H(\vec{d},n)^4He$ reaction on resonance at $E_d = 107$ keV at Basel [9] which was the occasion for the first polarization conference [1], the start of a long series of polarization symposia and high-energy spin conferences as well as many polarization workshops up to the present day (see e.g. [8]). The set-up of this experiment consisted of an atomic-beam polarized deuteron source connected to a *cascade generator* with 100 kV with a tritium target and plastic scintillator neutron detectors in the high-voltage dome. Figure 16.1 shows the experimental set-up and figures 16.2 and 16.3 show the construction details of the atomic-beam source. Under the assumptions that

- the reaction $^3H(\vec{d},n)^4He$ goes entirely through the resonant S-wave $J^\pi = 3/2^+$ matrix element of $^5He^*$,
- the ionization of the deuterium atoms takes place in a very weak magnetic field $B \to 0$, and
- no depolarization by unpolarized residual gas or non-adiabatic transitions occurs,

a vector polarization of $P_Z = 0$ and a tensor polarization of $P_{ZZ} = -1/3$ (in today's nomenclature) are predicted. The results of the measurements are shown in figure 16.4. The Erlangen source was designed for protons. The development after 1960 was rapid and efficient and can be best followed in the proceedings of numerous symposia, conferences and workshops on the subject. This first polarized source success set the pace for a rapid development of different types of polarized sources (and targets) with the aim of obtaining beam currents comparable to unpolarized beams and allowing experiments of high complexities such as spin-correlation and spin-transfer measurements in addition to simpler analyzing-power measurements. All these proved indispensable, e.g. for elucidating the details of the nucleon–nucleon interaction and the nature of the nuclear spin-dependent forces.

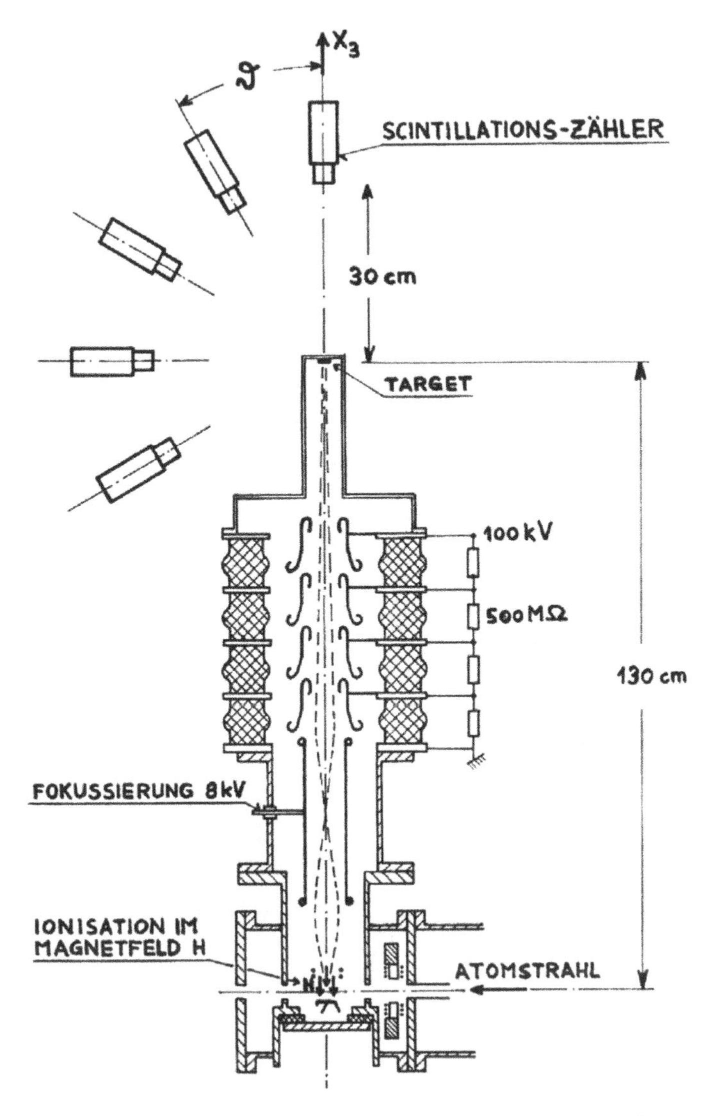

Figure 16.1. Schematic showing the first atomic-beam polarized ion source connected to an accelerator and used for the nuclear reaction $^3H(\vec{d},n)^4He$, thereby testing the sensitivity of this reaction as a polarization analyzer and checking the theoretical assumptions on the 107 keV resonance in ^5He as a pure S-wave resonance allowing no vector polarization sensitivity. Reproduced with permission after [9]. Copyright 1961 Schweizerische Physikalische Gesellschaft.

Later improvements of polarized ion sources consisted in:

- Replacement of the capillary collimators by a simple canal as the nozzle (which is possible with much increased (differential) pumping).
- Cooling of the nozzle, which leads to a higher beam density, higher acceptance of the Stern–Gerlach magnets and higher ionization efficiency.

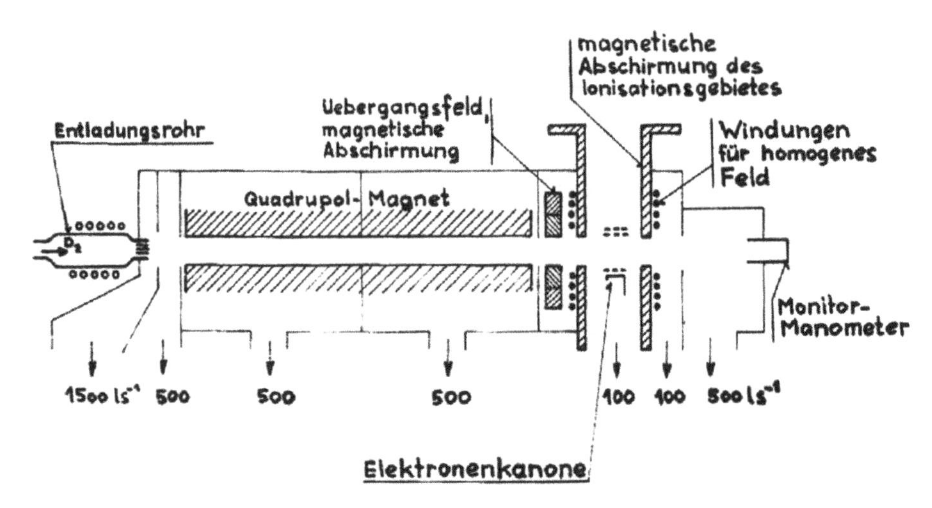

Figure 16.2. Construction of the atomic-beam polarized ion source using an RF-discharge dissociator, a glass-capillary collimator, long quadrupole magnets for spin-state separation and an electron-collision ionizer to produce a positive beam of partly tensor-polarized deuterons. Reproduced with permission after [9]. Copyright 1961 Schweizerische Physikalische Gesellschaft.

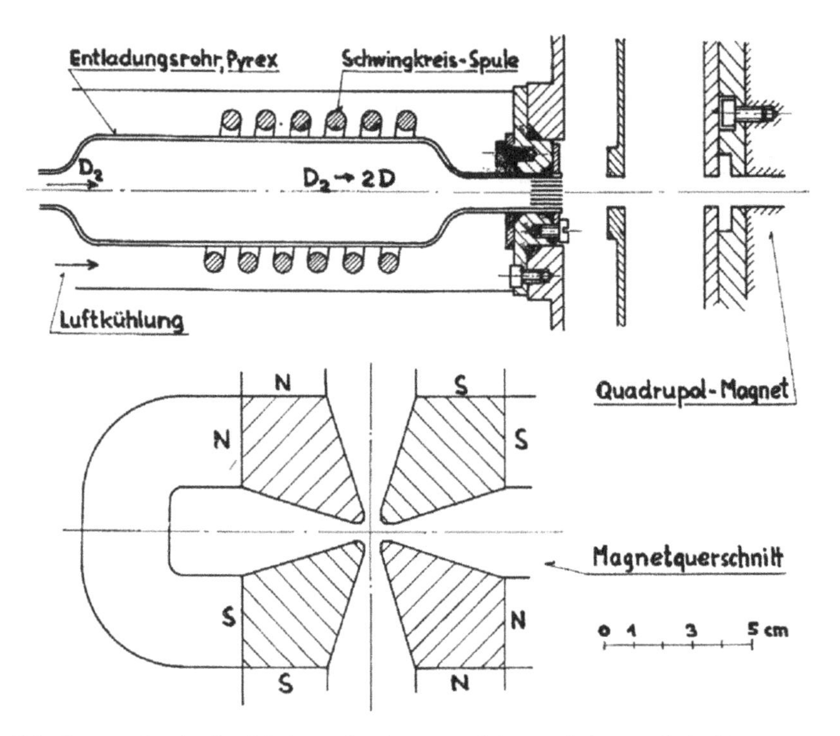

Figure 16.3. Construction details of the beam-forming parts of the atomic-beam polarized ion source using an RF-discharge dissociator, a glass-capillary collimator and long quadrupole magnets for spin-state separation. Reproduced with permission after [9]. Copyright 1961 Schweizerische Physikalische Gesellschaft.

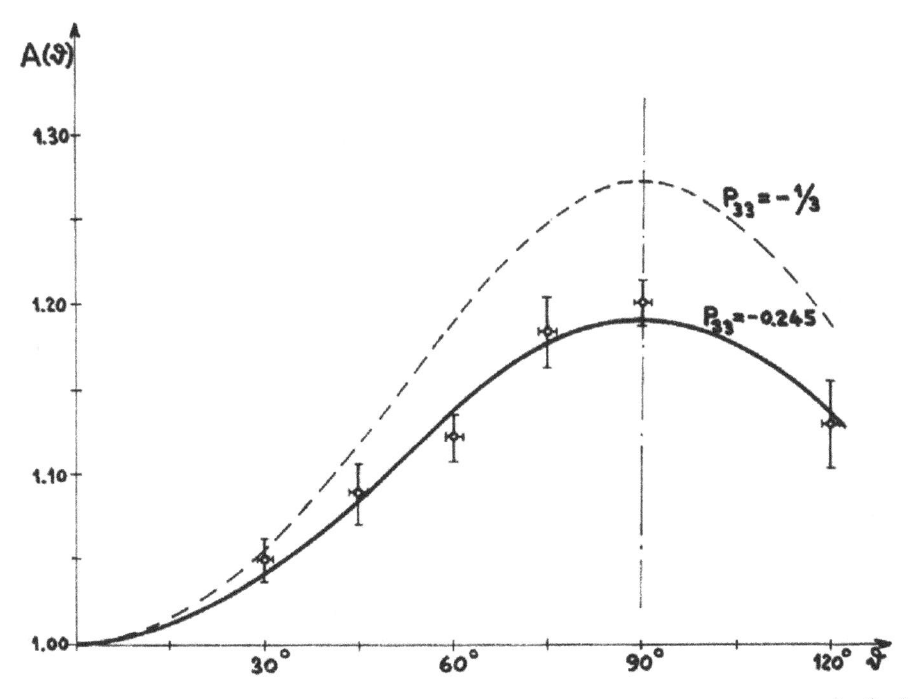

Figure 16.4. Angular distribution of the tensor analyzing power (here called $A(\theta)$) of the reaction $^{3}H(\vec{d},n)^{4}He$ measured at the first Basel source, compared to the prediction and with a best fit yielding $P_{ZZ} = -0.245$. Reproduced with permission from [9]. Copyright 1961 Schweizerische Physikalische Gesellschaft.

- Use of sextupoles (which have focusing properties on magnetic moments instead of quadrupoles that can only deflect them).
- RF transitions together with ionization in a strong magnetic field provide maximum values of polarization and high flexibility in changing the sign and also the type of polarization.
- Development of optimized ionizers (electron-collision or electron-cyclotron resonance type ionizers) as well as efficient charge-exchange methods for negative beams for tandem and other accelerators.
- The possibility to choose or handle the direction of the polarization vector at will, which is important because it enters directly into the planning and results of experiments. This is possible by using controlled precession of the polarization in magnetic fields, especially in Wien filters (crossed **E** × **B** devices).

More details on spin polarization in nuclear reactions and the necessary equipment and methods can be found, e.g., in [8].

An important point is to avoid *depolarization* of any form during the transport of the polarized beams to the target. Such depolarization may occur already in the ion source and during the stripping process in the stripper region of tandem Van de Graaff accelerators whenever the beam particles undergo a neutral phase with possible hyperfine interaction and non-adiabatic changes of magnetic fields. In cyclic accelerators such as storage rings (e.g. COSY-Jülich) *resonances* may cause depolarization. All these effects have been studied thoroughly and can be controlled.

Other types of sources have been developed based on the Lamb shift, the ionization by intense colliding particle beams such as Cs, or on optical pumping. Now typical are beams with polarizations of about 95% of the theoretical values and many tens of μA beam currents. A principal limitation to the intensity of the atomic beam lies in the gas dynamics around the dissociator/nozzle region ('intra-beam scattering'), in which sufficient pumping under a heavy gas load in tight spaces is limited.

Bibliography

[1] Huber P and Meyer K P (ed) 1961 *Proc. Int. Symp. on Polarization Phenomena of Nucleons* (*Basel 1960*) *Helv. Phys. Acta Suppl.* vol 6 (Basel: Birkhäuser)

[2] Clausnitzer G, Fleischmann R and Schopper H 1956 *Z. Phys.* **144** 336

[3] Clausnitzer G 1959 *Z. Phys.* **153** 600

[4] Dennison D M 1927 *Proc. R. Soc.* A **115** 483

[5] Frisch R and Stern O 1933 *Z. Phys.* **85** 4

[6] Gerlach W and Stern O 1920 *Z. Phys.* **9** 353

[7] Kapuscinski W and Eymers J G 1929 *Proc. R. Soc.* A **122** 58

[8] Paetz gen. Schieck H 2012 *Nuclear Physics with Polarized Particles* (*Lecture Notes in Physics* vol 842) (Heidelberg: Springer)

[9] Rudin H, Striebel H R, Baumgartner E, Brown L and Huber P 1961 *Helv. Phys. Acta* **34** 58

Chapter 17

The discovery of giant resonances

Giant resonances[1] are broad, resonance-like structures in excitation functions with large cross sections, excited by incident γ rays as well as in inelastic particle reactions such as (p,p′), (p,γ), and (α,α'). Their unusually large width is a consequence of the high excitation energy with many open decay channels and thus decay probabilities. The excitation energies of giant resonances as well as their widths follow a simple systematics and thus are similar between neighboring nuclei. Together with the large values of the cross sections this suggests a collective behavior of many (or all) nucleons.

Already in the years 1937–9 Walther Bothe and Wolfgang Gentner at Heidelberg investigated the nuclear photoeffect with γ rays from reactions such as ^{7}Li(p,γ)^{8}Be (17.2 MeV) and ^{11}B(p,γ)^{12}C (12.8 MeV). Using a 1 MV Van de Graaff accelerator ('Kanalstrahlanlage') they observed an increase of the (rather high) yield of the γ interaction with different target nuclei with energy and little dependence on the nuclide species. They remark that there seems to be no 'selective' behavior (i.e. resonance behavior) and nevertheless cite Bohr's compound-nucleus model for an explanation, but did not conceive of a (collective) giant-resonance phenomenon [10, 11].

With the advent of high-energy/high-intensity betatrons [21–23, 31] the upper energy edge of the continuous spectrum of bremsstrahlung γ rays was moved to >100 MeV. A number of experiments were performed after 1945, mainly to investigate different exit channels, and are summarized in [15]. The first experiments showing clear indications of such broad structures in different exit channels such as (γ,activation), (γ,fission), or (γ,n) were performed by Baldwin *et al* [1–3] using the 100 MeV betatron of the General Electric Laboratory at Schenectady, NY.

[1] It should be noted that in the older literature the term 'giant resonance' was also used for the total cross-section behavior of neutron scattering on many nuclei in an energy region where single resonances are not resolved and which is described by the optical model, see e.g. [26, 29, 30].

doi:10.1088/978-0-7503-1173-1ch17

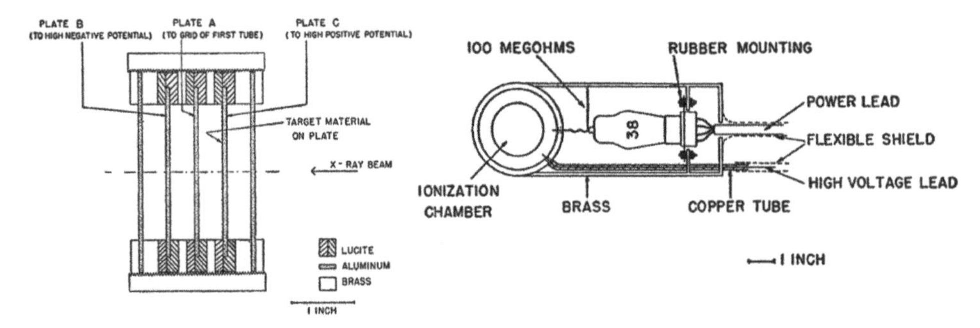

Figure 17.1. Construction details of the 'balanced' ionization chamber (left) and its housing (right). The 'target' substance was a coating on the inner electrode C. The other inner electrode B was uncoated. 'Balancing' of both units of the double ionization chamber was used to obtain clean fission signals in the presence of a strong x-ray background (a type of anti-coincidence technique). Adapted with permission from [2]. Copyright 1947 American Physical Society.

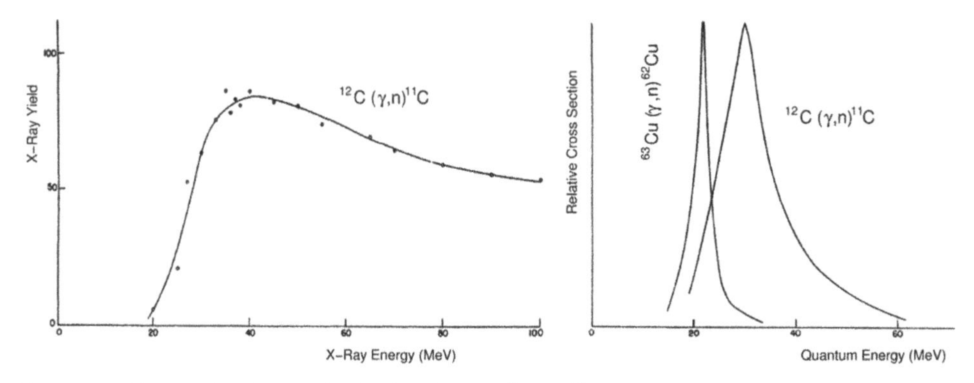

Figure 17.2. Photonuclear (γ,n) excitation functions on $^{12}C(\gamma,n)^{11}C$ before deconvolution of the bremsstrahlung spectrum produced by a betatron with $E_e = 100$ MeV (left) and deconvolved excitation functions of $^{12}C(\gamma,n)^{11}C$ and $^{63}Cu(\gamma,n)^{62}Cu$ (right), overlaid into one picture for comparison. Adapted with permission from [3]. Copyright 1948 American Physical Society.

Figure 17.1 shows the detection apparatus for photofission products in the presence of an intense x-ray background, as used by Baldwin *et al* The problem with bremsstrahlung excitation functions is that the measured intensities have to be deconvolved with the known bremsstrahlungs's spectrum. In all cases, for a number of target nuclei resonance-like excitation functions around energies of $\approx$20 MeV were found without explicitly calling the structures 'resonances' or even 'giant resonances' but discussing them in the framework of (statistical) compound-nucleus theories. One example of the results of these key experiments is shown in figure 17.2. The figures show the energy dependence and resonance widths of typical giant resonances. Similar results obtained for other exit channels (activation, fission) are evidence for the decays of a common intermediate state. With higher resolution fine structure of the giant-resonance peak has been found testifying to its character as an *intermediate structure/doorway* phenomenon.

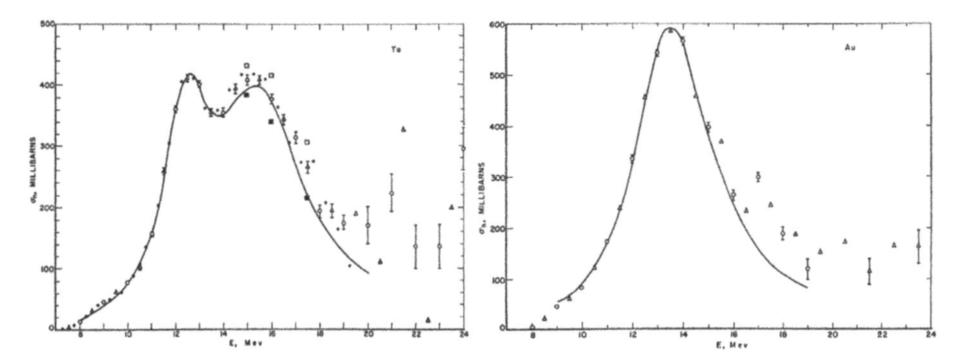

Figure 17.3. Total (γ,n) cross sections across the giant dipole resonances in tantalum and gold with a Breit-Wigner resonance shape fit. From the splitting of the giant dipole resonances the intrinsic quadrupole moment could be determined, e.g. for Ta: $Q_0 = 5.6 \pm 0.6$ b, corresponding to a prolate shape. Reproduced with permission from [17]. Copyright 1958 American Physical Society.

Based on the early data a theoretical description of the mechanism leading to the observed excitation functions was first given in 1948 by Goldhaber and Teller [18] assuming a resonant *dipole vibration* (an E1 excitation), i.e. a collective movement of the protons of nuclei against the neutrons. A semi-classical (hydrodynamic) model of the 'giant resonances' followed in 1950 from Steinwedel and Jensen [28], which delivered a reasonable account of the resonant energies over the periodic table. Improved measurements using bremsstrahlung were performed by Fuller *et al* [16, 17]. The problems of these measurements were still the deconvolution with the continuous spectrum and the difficulty of separating the (γ,2n) from the (γ,n) contribution. The methods used, however, were good enough to obtain the shape of the giant-resonance peak for many nuclei, also the double-hump structure of the peaks in the region of rare earths (reported independently by Spicer [27]), which had been predicted by Danos [13, 14] and simultaneously by Okamoto [24, 25]. Figure 17.3 shows two examples of the results of the experiments. The limitations of using bremsstrahlung γ rays are overcome by using photons with sharp energies. Already in 1937 Bothe and Gentner had used 17.4 MeV photons from the reaction ^{7}Li(p,γ) to investigate the nuclear photoeffect on many nuclei. Another method for experiments with giant-resonance studies developed by Bramblett *et al* [12] was to use nearly monochromatic annihilation γ from positrons produced and accelerated in a linear accelerator (LINAC) to up to 30 MeV. The LINAC had two stages: in the first electrons were accelerated to 10 MeV, then hit a W target to produce copious positrons from pair creation. These were accelerated to between 8–28 MeV producing annihilation photon pairs on a LiH target with variable energies. The photons in the forward direction obtained an energy of

$$h\nu = 1/2(2T_e + 3m_0c^2) = T_e + 0.77 \text{ MeV}$$

with T_e the kinetic energy of the positrons and m_0 the electron rest mass. The γ energy was measured using a NaJ(Tl) scintillation detector. Figure 17.4 shows the experimental arrangement. Figure 17.5 is an example of a (γ,n) excitation function

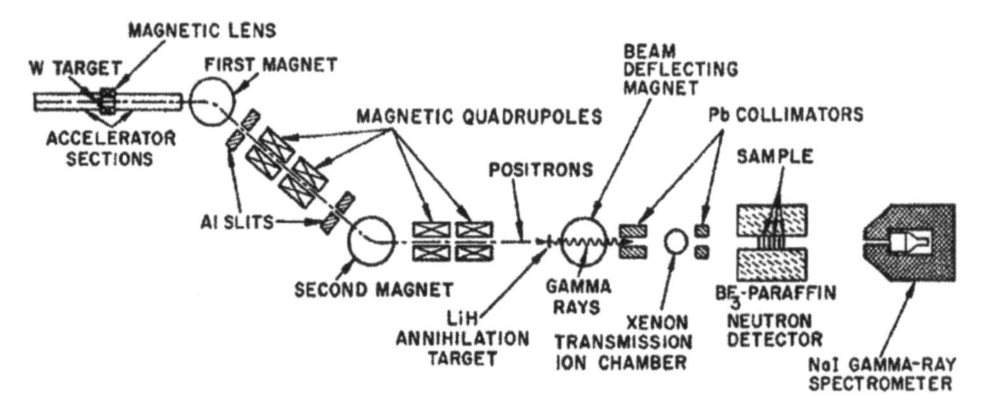

Figure 17.4. Set-up to produce γ rays via positron in-flight annihilation and detection of the (γ,n) neutrons in BF_3 detectors. Reproduced with permission from [12]. Copyright 1964 American Physical Society.

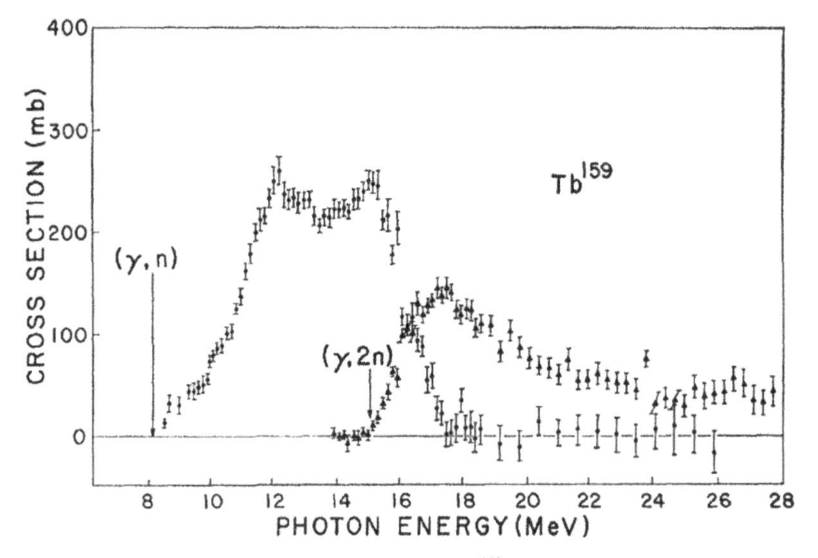

Figure 17.5. Measurement of the (γ,xn) excitation function for ^{159}Tb. Reproduced with permission from [12]. Copyright 1964 American Physical Society.

on ^{159}Tb showing the splitting of the giant resonance. Bramblett *et al* could deduce the intrinsic quadrupole moment

$$Q_0 = \left(7.0 \pm 1.1 \cdot 10^{-24}\right) \text{cm}^2$$

of the deformed nucleus ^{159}Tb from the ratio of the two peak energies, a value confirmed by Coulomb excitation. Figure 17.6 shows the systematic changes in the giant dipole resonance spectra of isotopes of Nd with increasing neutron number N, especially the shape change with the sudden onset of deformation with the isotope ^{148}Nd [4]. A relatively recent key experiment led to the discovery of the scissors-mode M1 giant resonance in ^{156}Gd [7]. Figure 17.7 shows a ^{156}Gd(e,e′) spectrum

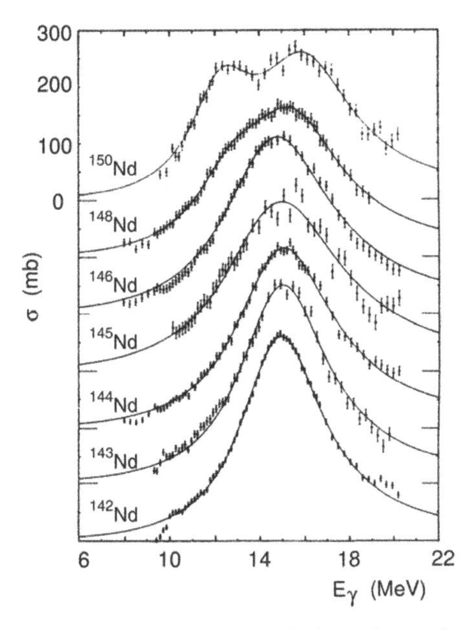

Figure 17.6. Giant dipole resonances in the photoneutron (γ,n) reaction on isotopes of Nd with increasing neutron number N, showing the onset of deformation. Adapted with permission from [4]. Copyright 1975 American Physical Society.

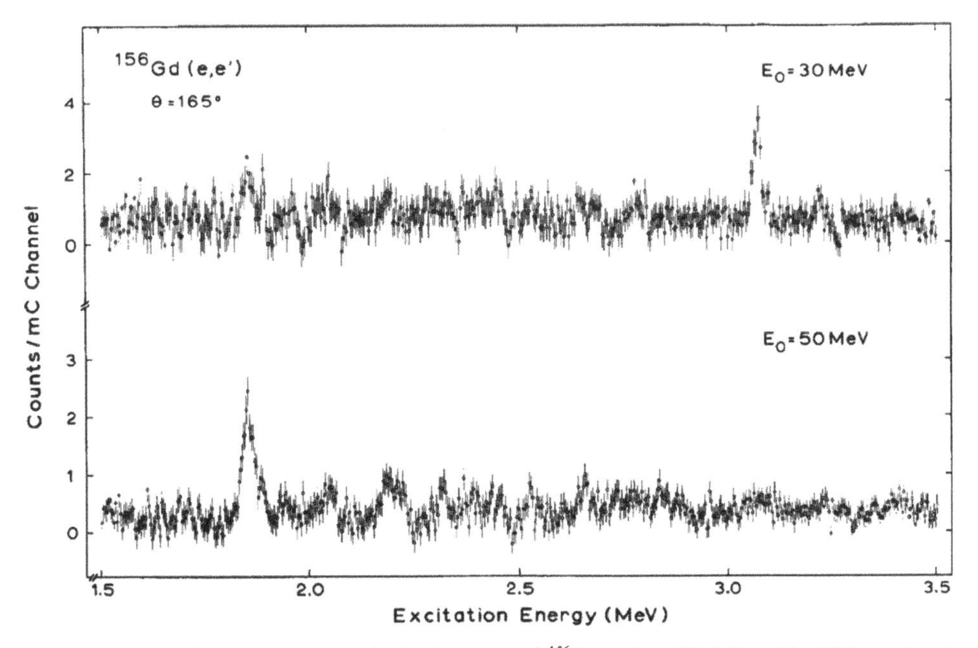

Figure 17.7. Giant dipole resonances in the (e,e′) reaction on ^{156}Gd at $E_0 = 30$ MeV and $\theta = 165°$ showing the newly found $J^{\pi} = 1^{+}$ (M1) resonance. At 50 MeV only a known 3^{-} state is visible. Reproduced with permission from [7]. Copyright 1984 Elsevier.

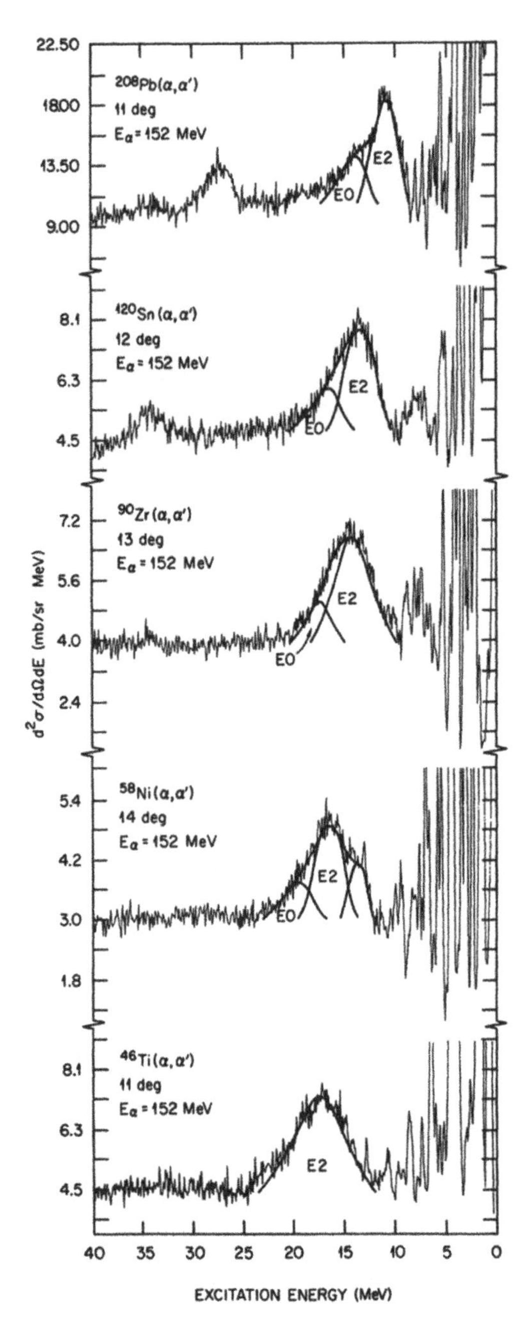

Figure 17.8. Giant resonances, e.g. monopole and quadrupole excitations, in inelastic α scattering (α, α') at $E_\alpha = 152\,\text{MeV}$. The high background makes the disentangling of different resonances difficult and may require model assumptions. Measurements are best performed at small forward angles. Reproduced with permission from [6]. Copyright 1980 American Physical Society.

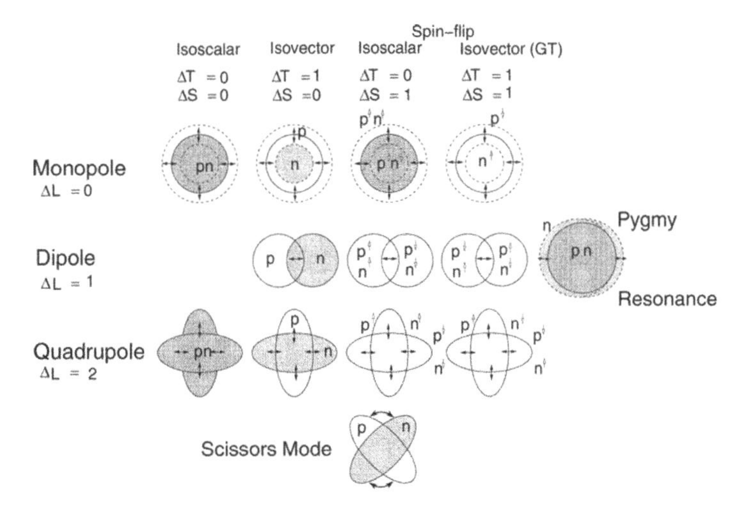

Figure 17.9. Giant resonances classified according to their multipolarity ΔL, their spin and isospin changes ΔS and ΔT.

obtained at the DALINAC electron accelerator. A special feature of this resonance is its narrow width and low excitation energy. It was found in many nuclei, especially in the rare-earth region, and corresponds to a scissor-like collective motion of the proton against the neutron liquid. Many giant resonance types have been found since. They can be classified according to their electromagnetic modes (or multipolarities), their isospin, or their motion types. These latter are

- The breathing mode: the entire nucleus 'breathes' without changing shape. Electromagnetically this is an E0 mode, its isospin is 0. It is plausible that this mode is related to the nuclear compressibility.
- The E1 mode: this is the classical dipole mode, in which proton and neutron fluids move collectively against each other.
- The M1, $T = 1$ mode: this is the magnetic-dipole scissors mode, in which the motions of the proton and neutron fluids have a rotational component acting like the arms of scissors.
- The E2 mode (or quadrupole giant resonance): protons and neutrons oscillate collectively against each other such that an oscillating quadrupole moment is formed.
- Higher-order resonances such as M3, E4, etc, have been found, e.g. the isoscalar octupole ($3\hbar\omega$) resonance.
- Relatively new developments concern the dipole (E1) pygmy resonance in nuclei with high neutron excess where—it is assumed—the neutron skin oscillates against the remaining $N = Z(T = 0)$ core. Radioactive ion beam facilities will be able to investigate this phenomenon near the borders of stability (driplines) in more detail.

The first-discovered giant resonances discussed above were of the 'classical' electric E1 dipole type.

Table 17.1. Properties of selected giant resonances.

Type of Resonance[a]	Character	ΔL	ΔS	ΔT	Excitation Energy E_x (MeV)	Width (MeV)	Preferred method	Relevance
ISGMR	E0	0	0	0 (IS)	$80A^{-1/3}$	3–5	(α,α')	Nuclear compressibility
IVGMR	E0	0		1 (IV)	$59A^{-1/6}$	10–15	Charge exch. $(\pi^{\pm},\pi^{0})$	
ISGDR	M1	1	1	0 (IS)				
IVGDR	E1	1	1	1 (IV)	$31.2A^{-1/3}$ $+20A^{-1/6}$	4–8	Photoabsorption	Historically first (1937)
ISGQR	E2	2	0	0 (IS)	$64.7A^{-1/3}$	$90A^{-2/3}$	(p,p′), (e,e′)	
IVGQR	E2	2	0	1 (IV)	$130A^{-1/3}$	5–15		
ISGOR (LEOR)	E3	3	0	0 (IS)	$41A^{-1/3}$			Low-energy OR
ISGOR (HEOR)	E3	3	0	0 (IS)	$108A^{-1/3}$	$140A^{-2/3}$		High-energy OR
GT	M1	0	1	1 (IV)	5–10		(p,p′), (p,n), (n,p)	β decay
Pygmy res.	E1	1	0	1 (IV)	9–12	2–4	(p,n), (^{3}He,t)	Neutron skin, Symmetry energy
Scissors mode	M1	1	1	1 (IV)	const $\approx$ 3		(e,e′)	Rotational component

[a] ISGMR: Isoscalar Giant Monopole Resonance, IVGMR: Isovector Giant Monopole Resonance, ISGDP: Isoscalar Giant Dipole Resonance, IVGDR: Isovector Giant Dipole Resonance, ISGQR: Isoscalar Giant Quadrupole Resonance, IVGQR: Isovector Giant Quadrupole Resonance, ISGOR: Isoscalar Giant Octupole Resonance, HEOR/LEOR: High Energy/Low Energy Octupole Resonance, GT: Gamow-Teller Resonance, IS/IV: Isoscalar/Isovector.

In addition to γ interactions, inelastic particle scattering, especially with α particles, is an important tool for measuring giant resonances. The cross sections peak at very small forward angles and the measurements normally require the use of magnetic spectrographs. Figure 17.8 shows an example of such spectra together with the indication of the disentanglement of the different contributions. Figure 17.9 shows schematically the classification of the different types of giant resonances. According to their collective character the resonance energies and widths vary systematically and slowly with mass number A. The energies of the resonance peaks are higher than those of 'normal' excitations, and the cross sections of the exciting reactions are high. This suggests the interpretation of the giant resonances as collective excitations of many (all) nucleons and—microscopically—the collective and coherent excitation of many-particle–many-hole (p–h) states across different harmonic-oscillator shells. As an example a collective $J^{\pi} = 1^{-}$ state may be constructed from excitation of $N = 5$ such p–h states from the p shell into the sd shell

$$\left|\left(1\mathrm{p}_{3/2}\right)^{-1}\left(1\mathrm{d}_{5/2}\right)\right\rangle_{1^-} \cdots \left|\left(1\mathrm{p}_{1/2}\right)^{-1}\left(1\mathrm{d}_{3/2}\right)\right\rangle_{1^-}. \tag{17.1}$$

A comprehensive survey of giant resonances up to 1999 is given in [19]. Inelastic scattering up to 1976 has been discussed in [5]. The more recent M1 scissors-mode giant resonance is discussed in detail in [20]. Some properties of important giant resonances are collected in table 17.1.

Bibliography

[1] Baldwin G C and Klaiber G S 1946 *Phys. Rev.* **70** 259
[2] Baldwin G C and Klaiber G S 1947 *Phys. Rev.* **71** 3
[3] Baldwin G C and Klaiber G S 1948 *Phys. Rev.* **73** 1156
[4] Berman B L and Fultz S C 1975 *Rev. Mod. Phys.* **47** 713
[5] Bertrand F E 1976 *Ann. Rev. Nucl. Part. Sci.* **26** 457
[6] Bertrand F E *et al* 1980 *Phys. Rev.* C **22** 1832
[7] Bohle W, Richter A, Steffen W, Dieperink A E L, Lo Iudice N, Palumbo F and Scholten O 1984 *Phys. Lett.* B **137** 27
[8] Bothe W and Geiger W 1925 *Z. Phys.* **32** 639
[9] Bothe W and Gentner W 1926 *Z. Physik* **35** 547
[10] Bothe W and Gentner W 1937 *Naturwissenschaften* **25** 90, 126 and 191
[11] Bothe W and Gentner W 1939 *Z. Phys.* **112** 45
[12] Bramblett R L, Caldwell J T, Harvey R R and Fultz S C 1964 *Phys. Rev.* B **133** 896
[13] Danos M 1956 *Bull. Am. Phys. Soc.* **1** 135
[14] Danos M 1958 *Nucl. Phys.* **5** 23
[15] Eyges L 1952 *Phys. Rev.* **86** 325
[16] Fuller E G, Petree B and Weiss M S 1958 *Phys. Rev.* **112** 554
[17] Fuller E G and Weiss M S 1958 *Phys. Rev.* **112** 560
[18] Goldhaber M and Teller E 1948 *Phys. Rev.* **74** 1046
[19] Harakeh M N and Van der Woude A 1999 *Giant Resonances—Fundamental High-Frequency Modes of Nuclear Excitations (Oxford Studies in Nuclear Physics* vol 24) (Oxford: Oxford Science)

[20] Heyde K, Richter A and von Neumann-Cosel P 2004 *Rev. Mod. Phys.* **82** 2365
[21] Kerst D W 1940 *Phys. Rev.* **58** 841
[22] Kerst D W 1941 *Phys. Rev.* **60** 47
[23] Kerst D W and Serber R 1941 *Phys. Rev.* **60** 53
[24] Okamoto K 1956 *Prog. Theor. Phys.* **15** 75
[25] Okamoto K 1958 *Phys. Rev.* **110** 143
[26] Satchler G R 1990 *Introduction to Nuclear Reactions* 2nd edn (London: McMillan)
[27] Spicer B M 1958 *Australian J. Phys.* **11** 298 and 490
[28] Steinwedel H, Jensen J H D and Jensen P 1950 *Phys. Rev.* **79** 1019
[29] Vogt E 1968 *The Statistical Theory of Nuclear Reactions* (*Advances in Nuclear Physics* vol 1) (New York: Plenum) p 261
[30] Vogt E 1972 *Rev. Mod. Phys.* **34** 723
[31] Wideröe R 1928 *Arch. Elektrotech.* **21** 387

Lightning Source UK Ltd.
Milton Keynes UK
UKOW07n1937031215

264068UK00001B/2/P